Meaning and Motoricity

Kristóf Nyíri

# Meaning and Motoricity
Essays on Image and Time

**Bibliographic Information published by the Deutsche Nationalbibliothek**
The Deutsche Nationalbibliothek lists this publication in the Deutsche Nationalbibliografie; detailed bibliographic data is available in the internet at http://dnb.d-nb.de.

**Library of Congress Cataloging-in-Publication Data**
Nyíri, János Kristóf.
 Meaning and motoricity : essays on image and time / Nyíri, Kristóf. - 1 [edition].
   pages cm
 Includes index.
 ISBN 978-3-631-65134-6
 1. Image (Philosophy) 2. Time–Philosophy. I. Title.
 B105.I47N95 2014
 128'.4–dc23
                                                2014007733


Cover illustration:
Albrecht Dürer's woodcut "Der Tod und der Landsknecht" (1510).
Image by courtesy and kind permission of the
Herzog Anton Ulrich-Museum Braunschweig,
Kunstmuseum des Landes Niedersachsen.


ISBN 978-3-631-65134-6 (Print)
E-ISBN 978-3-653-04266-5 (E-Book)
DOI 10.3726/978-3-653-04266-5

© Peter Lang GmbH
Internationaler Verlag der Wissenschaften
Frankfurt am Main 2014
All rights reserved.
Peter Lang Edition is an Imprint of Peter Lang GmbH.

Peter Lang – Frankfurt am Main · Bern · Bruxelles · New York ·
Oxford · Warszawa · Wien

All parts of this publication are protected by copyright. Any utilisation outside the strict limits of the copyright law, without the permission of the publisher, is forbidden and liable to prosecution. This applies in particular to reproductions, translations, microfilming, and storage and processing in electronic retrieval systems.

www.peterlang.com

# Contents

Preface ........................................................................................................... 7

1. Visualization and the Horizons of Scientific Realism ........................ 11

2. Hundred Years After:
   How McTaggart Became a Thing of the Past .................................... 35

3. Gombrich on Image and Time ............................................................ 53

4. Image and Metaphor in the Philosophy of Wittgenstein .................... 73

5. Time As a Figure of Thought and As Reality ..................................... 93

6. Images in Conservative Education .................................................... 105

7. Time and Image in the Theory of Gestures ....................................... 121

Index .......................................................................................................... 145

# Preface

Pictures and pictorial meaning did rarely become philosophical topics before the twentieth century. The reason has quite clearly to do with technology, namely with the technology of communication. Prior to 1400, European culture was not familiar with any technologies for duplicating pictures, an exact pictorial representation of reality was impossible before the age of photography, to deal with images was much more cumbersome than to deal with texts, philosophers communicated in words about words. In the twentieth century however there emerged, within a few decades, satisfactory answers to the fundamental questions of the philosophy of images – answers, to be sure, still today largely rejected by the philosophical community. I believe the definitive work done here is that by Ernst Gombrich. The journey he travelled from the 1960s to the end of the 1970s is telling. In his *Art and Illusion* (1960) he highlighted the role of conventions in pictorial representation. In his 1969 paper "The Evidence of Images" he still stressed that images without words are not unequivocal: discussing Dürer's woodcut "Death and the Landsknecht" (1510) he pointed out that here the artist himself seems to have felt necessary to support the pictorial message by a rhymed text – "Vnd thu stetz noch gnaden werben/Als soltestu all stund sterben" ("Always seek for grace/As if you might die any moment"). By 1978 however, in his essay "Image and Code", Gombrich came to argue for the idea that images might be self-evident natural signs.

By contrast, since the two fundamental pronouncements of Aristotle – "time is the number of movement in respect of the before and after", yet it is "a question that may fairly be asked … [w]hether if soul did not exist time would exist or not" (*Physics*, 220a25, 223a22–23, Hardie–Gaye transl.) – the problem of time has, for all the genius of Bergson, Heidegger, or indeed Einstein, apparently not come any closer to a solution. In my view the reason for this is that any appropriate philosophy of time will presuppose an appropriate philosophy of images. Time and image refer to each other, and in particular it is not possible to build up an argument for the reality of the passage of time without accepting that pictorial meaning is essentially non-conventional, images being expressions of physical forces acting on us. This volume represents a rudimentary attempt towards such an argumentation. I am aware of going against the stream both when it comes to the philosophy of images and the philosophy of time – drawing courage not so much from philosophy, as rather, say, from Russian film director Andrey Tarkovsky's notion of "time-pressure" and his idea of a "time flowing" with "dignity, independently" (*Sculpting in Time*, transl. by Kitty Hunter-Blair, University of Texas Press, 1987, pp. 117 and 120).

The first chapter of the volume serves to introduce my main topics. Verbal thinking, as also mathematical thinking, is fundamentally intertwined with, and indeed presupposes, visual thinking, while all involve an underlying motor dimension. The significance of the visual was recognized both by Heidegger and Wittgenstein, but still went mostly unnoticed in twentieth-century philosophy, with the result that defenders of scientific realism in the philosophy of science, most importantly perhaps Wilfrid Sellars, were not capable of exploiting an important line of argument indispensable to their position. In particular, in the philosophy of time the purported realism of "four-dimensionalism" will be seen as phoney once visual imaginability is accepted as a criterion of intelligibility. Four-dimensionalism is often taken to be related to McTaggart's "B-theory". In my second chapter, I endeavour to show the spuriousness of McTaggart's arguments, suggesting that their baffling popularity might well have to do with the parallel appearance of, and a mistaken similarity to, the Einstein–Minkowski conception of space-time. I conclude the chapter with a first brief interim summary of the way I believe a philosophical argument for the vindication of the common-sense view of time might proceed.

The third chapter is a survey of Gombrich's writings on pictorial meaning, on the interdependence of word and image, on how movement can be suggested by static images, and on how the passage of time is represented by pictures themselves immobile. Gombrich takes issue with the notion of a *punctum temporis*, of static points of time (inevitably leading, as he emphasizes, to Zeno's paradox), arguing for the idea of the "specious present", a broader time span present to the mind, a time span that allows for the immediate perception of real change.

As I indicate throughout this volume, to Gombrich's work a felicitous and indeed necessary complement is that of Rudolf Arnheim. I am discussing Arnheim in some detail in chapters 5 and 6, but, before that, I have to present an admittedly iconoclastic, and hopefully convincing, image of my life-long hero, Ludwig Wittgenstein. I do this in chapter 4 (coming then back to Wittgenstein again in chapter 6). Wittgenstein, I submit, was a precursor of the iconic turn, while of course being one of the main actors of the foregoing linguistic one. However, to his and to our detriment, he never succeeded in synthesizing his views on language on the one hand and images on the other, and was crucially unsuccessful precisely when it came to developing the theory in which words and pictures should from the very beginning meet: metaphor theory.

No adequate philosophy of time is possible without an adequate theory of metaphors. I attempt to establish the connection between the two domains in chapter 5. And it is in this chapter I actually try to sum up my argument for the position that the common-sense view of the reality of time is philosophically defensible. Chapters 6 and 7 adduce further elements to this argument. In chapter 6, "Images in Conservative Education", I emphasize the capacity of the visual mind to mirror physical reality, suggest that today's abundance of informative and indeed veridical images, still and moving, redeem us from a more or less uncritical reliance on often delusive texts, from verbal traditions handed down by word of mouth or in writing, and that the diminishing role of traditions implies a changed sense of past, present and future. Chapter 7 focusses on bodily gestures as translating the motor into the visual, and concludes, once more, that the experience of the passage of time is an embodied, primary one. The section "Meaning and Motoricity" in chapter 7 provides a brief summary of the ideas which made me conceive of the title of the present volume.

The essays here collected, written in the course of the past six years or so, have, with the exception of the last one, already appeared in print, but they were from the very beginning meant to become chapters of a single book. Chapter 1, "Visualization and the Horizons of Scientific Realism", emerged from a talk given in 2008 at a conference in Pécs (Hungary) on Richard Rorty, and was, in an extended form, published in András Benedek and Kristóf Nyíri (eds.), *The Iconic Turn in Education* (series VISUAL LEARNING, vol. 2), Frankfurt: Peter Lang, 2012. Chapter 2, "Hundred Years After: How McTaggart Became a Thing of the Past", appeared in T. Czarnecki et al. (eds.), *The Analytical Way: Proceedings of the 6th European Congress of Analytic Philosophy*, London: College Publications, 2010. Chapter 3, "Gombrich on Image and Time", was published online in the *Journal of Art Historiography*, no. 1 (December 2009), and in hardcopy in Klaus Sachs-Hombach and Rainer Totzke (eds.), *Bilder – Sehen – Denken: Zum Verhältnis von begrifflich-philosophischen und empirisch-psychologischen Ansätzen in der bildwissenschaftlichen Forschung*, Köln: Herbert von Halem Verlag, 2011. Chapter 4, "Image and Metaphor in the Philosophy of Wittgenstein", appeared in R. Heinrich et al. (eds.), *Image and Imaging in Philosophy, Science and the Arts*, Proceedings of the 33rd International Ludwig Wittgenstein Symposium, vol. 1, Heusenstamm bei Frankfurt: ontos verlag, 2011. Chapter 5, "Time As a Figure of Thought and As Reality", was published in András Benedek and Kristóf Nyíri (eds.), *Images in Language: Metaphors and Metamorphoses* (series VISUAL LEARNING, vol. 1), Frankfurt: Peter Lang, 2011. Chapter 6, "Images in Conservative Education", appeared in András Benedek and Kristóf Nyíri (eds.),

*How to Do Things with Pictures* (series VISUAL LEARNING, vol. 3), Frankfurt: Peter Lang, 2013. I gratefully acknowledge the permission of Peter Lang Verlag, and of András Benedek, co-editor of the series VISUAL LEARNING, to reprint Chapters 1, 5, and 6. As indicated above, chapter 7, "Time and Image in the Theory of Gestures", is here published for the first time.

Dunabogdány, March 2014

# 1. Visualization and the Horizons of Scientific Realism

Galileo's often-quoted formula, according to which the universe is written in the language of mathematics, continues with the elucidation, "its characters are triangles, circles, and other geometrical figures, without which it is humanly impossible to understand a single word of it".[1] Above the doorway of Plato's Academy, some two millennia earlier, there was engraved, as tradition has it, the inscription "Let no-one ignorant of geometry enter here" – rendered not infrequently, and not without justification, as "Let no one ignorant of mathematics enter here": since for the Greeks it was precisely geometry that constituted the essence of mathematics. And for Plato in a sense all branches of mathematics, and indeed all branches of thought, had to do with shapes. He chose the words *idea* and *eidos* to designate abstract mental contents. These words, which he used alternately, mean "form" or "shape". Both *idea* and *eidos* come from the verb *idein*, "to see"; from *eidos* there descends the word *eidolon*, "the visible image".[2] In the writings of Archimedes and Apollonius *eidos*, along with *schēma*, again with the meaning "figure" or "shape", emerged as parts of the mathematical lexicon.[3] History shows mathematics to be inherently bound up with visuality. In fact any dimension of abstract reasoning does essentially rely on the perceptual, in particular on the visual: mental processes invariably involve the component of imagery.

## The Visual Mind

As a fairly recent, succinct summary by Kosslyn et al. puts it: "Mental imagery occurs when perceptual information is accessed from memory, giving rise to the

---

1 Cf. James Franklin, "Diagrammatic Reasoning and Modelling in the Imagination: The Secret Weapons of the Scientific Revolution", in Guy Freeland and Anthony Corones (eds.), *1543 and All That: Image and Word, Change and Continuity in the Proto-Scientific Revolution*, Dordrecht: Kluwer, 2000, pp. 53 f.
2 I am indebted to István Bodnár for innumerable enlightening conversations, in the course of the years, on some intricate issues in Greek philosophical usage and intellectual history.
3 Cf. Reviel Netz, *The Shaping of Deduction in Greek Mathematics: A Study in Cognitive History*, Cambridge: Cambridge University Press, 1999, pp. 109 f.

experience of 'seeing with the mind's eye', 'hearing with the mind's ear' and so on. ... Mental images need not result simply from the recall of previously perceived objects or events; they can also be created by combining and modifying stored perceptual information in novel ways. Imagery has had a central role in theories of mental function since at least the time of Plato."[4] Now Plato's views on mental images are of course deeply ambiguous. His philosophy emerged under the impact of the rise of *alphabetic literacy*. And, from Plato onwards, the history of Western philosophy is a history of recurrent clashes between the experience of imagery on the one hand, and the experience of written language on the other.

## From Plato to Hume

Though *eidos* is not etymologically related to *eikon* – "likeness", "picture" – the acoustic and semantic proximity between the two words does suggest a kind of relatedness, and Plato is not always willing, or able, to avoid that suggestion. But in the *Phaedrus* he definitely tells us that "essences" are "formless, colourless, intangible, perceived by the mind only",[5] and in the *Republic* we learn that "ideas can be thought but not seen"[6]. Also, in a telling passage of the *Philebus* Plato compares the soul to a *book*, adding however that besides the "scribe" who writes "within us" there is also "another artist, who is busy at the same time in the chambers of the soul": "The painter, who, after the scribe has done his work, draws images in the soul of the things which he has described."[7] Aristotle's *De anima* is dominated by the metaphor of the mind as a "writing-table" (*grammateion*), but still it is here that the momentous thesis is formulated according to which "the soul never thinks without an image" (*phantasma*).[8]

It was on the teachings of the Aristotelian school that Bacon drew when he wrote: "Emblem reduceth conceits intellectual to images sensible, which strike the memory more. ... Aristotle saith well, 'Words are the images of cogitations, and letters are the images of words.' But yet it is not of necessity that cogitations

---

4   Stephen M. Kosslyn – Giorgio Ganis – William L. Thompson, "Neural Foundations of Imagery", *Nature Reviews Neuroscience*, vol. 2, no. 9 (2001), p. 634.
5   247c, Jowett transl.
6   507b, Shorey transl.
7   39a-b, Jowett transl.
8   Aristotle, *On the Soul*, 430a and 431a, transl. by J. A. Smith. *The Complete Works of Aristotle: The Revised Oxford Translation*, ed. by Jonathan Barnes, Princeton: Princeton University Press, 1984.

be expressed by the medium of words. For whatsoever is capable of sufficient differences, and those perceptible by the sense, is in nature competent to express cogitations."[9] By contrast, Descartes asks us to "recall that our mind can be stimulated by many things other than images – by signs and words, for example, which in no way resemble the things they signify", pointing out, also, that "the perfection of an image often depends on its not resembling its object as much as it might".[10] The British empiricist reaction to Descartes is again characterized by an enhanced sensibility to images, with Locke however retaining a conspicuous susceptibility to the lure of written language. On the one hand Locke, very much in the spirit of Bacon, reflects on the advantages of a dictionary in which "words standing for things which are known and distinguished by their outward shapes [w]ould be expressed by little draughts and prints made of them".[11] And he equates – albeit not always unambiguously – ideas with mental images, for instance in the section on *Abstraction*, where he says: "ideas taken from particular beings become general representatives of all of the same kind; and their names general names, applicable to whatever exists conformable to such abstract ideas. Such precise, naked appearances in the mind ... the understanding lays up (with names commonly annexed to them) as the standards to rank real existences into sorts."[12] The words "idea", "conception", "thought" and "imagination" Locke tends to treat as synonymous.[13] On the other hand in the *Essay* there is a marked tendency to equate ideas with single *written words*. The mind, at birth, is like a "white paper, void of all characters, without any ideas"; when describing the doctrine of *stamped*, or *imprinted*, innate characters,[14] it is only the innateness Locke takes issue with. For Berkeley and Hume it was not at all a question that ideas are mental images; their problem, rather, was to understand how images can be the carriers of *general meanings*.[15]

---

9   Francis Bacon, *The Advancement of Learning* (1605), Oxford: Clarendon Press, 1974, pp. 130 f. Bacon's reference here is to the *De interpretatione*.
10  *The Philosophical Writings of Descartes*, vol. I, Cambridge: Cambridge University Press, 1985, p. 165.
11  John Locke, *An Essay Concerning Human Understanding*, Book III, ch. xi, sect. 25.
12  *Ibid.*, Book II, ch. xi, sect. 9.
13  Cf. e.g. Book III, ch. ii, sect. 6.
14  *Ibid.*, Book II, ch. i, sect. 2; and Book I, ch. i, sect. 1 and 5.
15  Recall Locke's famous difficulty, described in Book IV, ch. vii, sect. 9 of his *Essay*. As Locke here puts it, it does indeed "require some pains and skill to form the general idea of a triangle, (which is yet none of the most abstract, comprehensive, and difficult,) for it must be neither oblique nor rectangle, neither equilateral, equicrural, nor scalenon; but all and none of these at once. In effect, it is something imperfect, that

Berkeley, insisting that ideas are indeed images, maintained that generic mental images are inconceivable. Hume however seems to have found a solution: according to his formula in the section "Of Abstract Ideas" of Book One in the *Treatise*, we have ideas "not really and in fact present to the mind, but only in power", ideas we do not "draw ... all out distinctly in the imagination, but keep ourselves in a readiness to survey any of them, as we may be prompted by a present design or necessity". Of the British Empiricists, it is Hume whose views on the thinking process are the most consistently imagistic.

## The Darwin Effect

The British Empiricists' perceptivity for the role of mental images left no trace on the philosophical thinking of the last decades of the 18th and the first half of the 19th centuries. Kant's heroic attempt, in the chapter on schematism in his *Critique of Pure Reason*, to come to terms with the problem of how images and concepts hang together, had absolutely no impact for some hundred and fifty years;[16] philosophy, both on the Continent and in Britain, became for quite some time entirely language-centred. The "linguistic turn", to use the expression made famous by Richard Rorty whose views I will briefly discuss further below, much pre-dated the twentieth century. By way of illustration, let me here give three indirect references. First, the neurologist Henry Head rebelling, in the 1920s, against a very influential paper by Bastian, published in 1869, with Head remarking that "the whole work" of Bastian "was founded on the axiom that 'we think in words'".[17] Secondly, Ribot taking issue, in 1897, with Max Müller, the German-born philologist and orientalist of great renown, working in Britain. As Ribot puts it, Müller accepts as

---

cannot exist; an idea wherein some parts of several different and inconsistent ideas are put together". Ideas seem to be of a pictorial nature (otherwise the general idea of a triangle would not cause embarrassment) but also they must permit of non-pictorial dimensions (since as *generic* pictures, Locke implies, they *cannot* exist).

16  Heidegger's *Kant and the Problem of Metaphysics*, the first study to discover the significance of these Kantian analyses, appeared in 1929. See my paper "Kritik des reinen Bildes: Anschauung, Begriff, Schema", in H. Lenk and R. Wiehl (eds.), *Kant Today/Kant aujourd'hui/Kant heute*, Münster: LIT, 2006, pp. 71–84, and my discussion of Heidegger in the present chapter.

17  Henry Head, *Aphasia and Kindred Disorders of Speech*, Cambridge: Cambridge University Press, 1926, vol. 1, p. 54, referring to H. Charlton Bastian, "On the Various Forms of Loss of Speech in Cerebral Disease". Head is arguing for a "return to the teaching of Aristotle that human reason depends on the senses and imagery", *ibid.*, p. 45.

an "axiom" the "antique aphorism" according to which "it is impossible to think without words".[18] And thirdly: half a century later the mathematician Hadamard is still outraged by Müller, who "claims to find in the fact that thought is impossible without words an argument against every evolutionary theory, a proof that man cannot be descended from any animal species".[19]

That evolutionary theory should enter the picture here is significant. One can witness a late-nineteenth-century revival of the interest in mental images, an early, and for decades forgotten, prelude to the iconic turn proper beginning in the 1970s, an interest that was unequivocally bound up with the impact of Darwin. Before Darwin, there was reason to take the abyss between animal and human intelligence for granted: animal mental life might be based on images, but that of humans was based on language ("in the beginning was the word"). With Darwin this changed. The *Descent of Man* speaks a clear language: "The *Imagination* is one of the highest prerogatives of man. By this faculty he unites former images and ideas, independently of the will, and thus creates brilliant and novel results. ... Dreaming gives us the best notion of this power... As dogs, cats, horses, and probably all the higher animals, even birds have vivid dreams, and this is shewn by their movements and the sounds uttered, we must admit that they possess some power of imagination."[20]

It is not by chance that it was Darwin's early advocate T. H. Huxley who in his book on Hume, published in 1878, ventured to return to the topic of "generic ideas" that "may exist independently of language", ideas which Huxley compares to "compound photographs", amounting to "sketches", *generic portraits*, rather than a specific portrait.[21] And it was Darwin's half-cousin Galton who in his *Inquiries into Human Faculty*, published in 1883,[22] first outlined,

---

18  Théodule Armand Ribot, *L'évolution des idées générales*. I am quoting from the English translation, *The Evolution of General Ideas*, Chicago: Open Court, 1899, p. 39, see also p. 28: "Max Müller, who persists in affirming that it is radically impossible to think and reason without words…"

19  Jacques Hadamard, *An Essay on the Psychology of Invention in the Mathematical Field*, Princeton, NJ: Princeton University Press, 1945, p. 67.

20  Charles Darwin, *The Descent of Man, and Selection in Relation to Sex*, vol. 1, London: John Murray, 1871, pp. 45 f.

21  There are very many editions of Huxley's *Hume*; a convenient summary of his views on the subject of imagery is provided by William James, in his *The Principles of Psychology* (1890), in the chapter on "Imagination", London: Macmillan & Co., 1901, vol. II, pp. 46 ff.

22  Francis Galton, *Inquiries into Human Faculty and Its Development* (1883), 2nd ed. London: J. M. Dent & Co., 1907.

based on empirical investigations, a well-rounded and extremely influential theory of mental images, a theory with immediate impact on Binet,[23] James[24] and Ribot,[25] and exploited somewhat later by Titchener,[26] Koffka,[27] Russell[28] and innumerable others, with echoes even in Wittgenstein's thinking.[29] One of the addressees of the questionnaires Galton has sent out was Darwin himself. In his reply, Darwin, as Howard Gruber puts it, "gives an account of himself as someone with fairly strong visual imagery";[30] his answer to the question as to whether he has, and what kind of, visual recollections of his breakfast table, runs: "Some objects quite defined, a slice of cold beef, some grapes and a pear, the state of my plate when I had finished, and a few other objects, are as distinct as if I had photos before me."[31] Indeed *thinking with* mental images, and thinking with *diagrams* emerging from mental images, seems to have been, as Gruber has shown, a centrally important method for Darwin. The "tree of

---

23  Cf. Alfred Binet, *La psychologie du raisonnement* (1886), Engl. translation *The Psychology of Reasoning*, Chicago: Open Court, 1899, pp. 25 f. and 116–118.
24  Cf. *op. cit.*, vol. 2, pp. 51 ff.
25  Cf. Ribot, *op. cit.*, p. 10.
26  Cf. Edward Bradford Titchener, *Lectures on the Experimental Psychology of the Thought-Processes*, New York: Macmillan, 1909, pp. 13, 201 f., 205 f., 208, 211.
27  Cf. Kurt Koffka, *Zur Analyse der Vorstellungen und ihrer Gesetze: Eine experimentelle Untersuchung*, Leipzig: Quelle & Meyer, 1912, p. 194.
28  See Bertrand Russell, "On Propositions: What They Are and How They Mean" (1919), *Aristotelian Society Supplementary Volume*, 2, pp. 1–43, repr. in J. G. Slater (ed.), *The Collected Papers of Bertrand Russell*, Volume 8: *The Philosophy of Logical Atomism and Other Essays, 1914–19*, London: George Allen & Unwin, 1986, pp. 284 f.: "If you try to persuade an ordinary uneducated person that she cannot call up a visual picture of a friend sitting in a chair, but can only use words describing what such an occurrence would be like, she will conclude that you are mad. (This statement is based upon experiment.) I see no reason whatever to reject the conclusion originally suggested by Galton's investigations, namely, that the habit of abstract pursuits makes learned men much inferior to the average in the power of visualizing, and much more exclusively occupied with words in their 'thinking'."
29  Cf. Ludwig Wittgenstein, *Preliminary Studies for the "Philosophical Investigations": Generally Known as the Blue and Brown Books*, ed. by Rush Rhees (1958), Oxford: Basil Blackwell, 1964, p. 18.
30  Howard E. Gruber, *Darwin on Man: A Psychological Study of Scientific Creativity*, 2nd ed., Chicago: The University of Chicago Press, 1981, p. 237. I am indebted to Csaba Pléh for having drawn my attention to Gruber's work.
31  *The Life and Letters of Charles Darwin*, vol. III, London: John Murray, 1887, p. 239.

life" diagram, published in the *Origins of Species*, has a number of forerunners in Darwin's notebooks – the "tree schema" there actually serves as a basis of specific deductions.[32]

## Meeting Rorty

The thesis I am arguing for in this opening chapter is that the demarcation line beyond which we should conceive of scientific theories not as possible explanations of the world, but as mathematical instruments enabling us to arrive at correct practical predictions, is not the much-discussed observable/non-observable border (in the case of Mach and the logical positivists blending into the demarcation line between science and metaphysics), but rather the border between, on the one hand, what we can *imagine*, in the sense of being able to form perceptual images, and, on the other hand, what we cannot describe but in abstract symbolic terms. The present section and the next one are meant to set the stage, in the form of some personal reminiscences and reflections, for my main argument which I will present in the last, somewhat longer, section: "Believe What You Can Visualize".

I became personally acquainted with Rorty late in both of our lives. I met him for the first and the last time in 2004, on two consecutive days. On May 5 I picked him up, with his wife, at the railway station in Budapest where they arrived from a visit in Pécs in southern Hungary. I drove them to their hotel and we discussed some organizational details in connection with the talk he was to give on the next day at the Hungarian Academy of Sciences. He seemed tired; we soon parted. I vividly remember the following morning. There was still some time before his talk was due, the sun was shining beautifully; we walked a short distance from the Academy main building to the Danube – to the Chain Bridge – and suddenly I found myself asking him a question. What did he think, I asked, about the *pictorial turn* underway in philosophy? Clearly, this was a rather extraordinary question to put to the man whose name had been, ever since the mid-1960s, closely associated with the term "linguistic turn"[33], and whose 1979 book *Philosophy and the Mirror*

---

32  Cf. Gruber, *op. cit.*, pp. 141–144, see also Howard E. Gruber, "Darwin's 'Tree of Nature' and Other Images of Wide Scope" (1978), in Howard E. Gruber – Katja Bödeker (eds.), *Creativity, Psychology and the History of Science*, Dordrecht: Springer, 2005, pp. 241–257.

33  The term itself Rorty attributes to Hugo Bergmann, cf. the editor's "Introduction" in Richard Rorty (ed.), *The Linguistic Turn. Recent Essays in Philosophical Method*, Chicago: The University of Chicago Press, 1967, p. 9.

*of Nature* was a single extended attack on "ocular" or "visual" metaphors in philosophy[34] – on the "spectator theory of knowledge"[35]. But wasn't W. J. T. Mitchell's 1992 paper "The Pictorial Turn" directly addressing Rorty's work,[36] and didn't the latter by 1990 regard the issues pertaining to linguistic philosophy as having become quaint? Rorty's reaction, there and then, was embarrassing: he has never heard about the expression "pictorial turn", could not imagine what it might mean, and was utterly taken aback by my hurried attempt at some rudimentary explanation. Still, the subject came up again later in the day, during the dinner to which I invited the couple at a restaurant in my home village on the Danube Bend. I think I tried to say something about the implications, for philosophy, of the imagery debate in cognitive science, and about how the ease of accessing and indeed producing pictures in the new digital medium affects not only the ways we communicate, but also the ways we think. This time Dick became interested, as did, also, Mary; they were empathetic, inspiring, and of course absolutely charming; we decided that we should stay in touch and continue discussing the topic.

It did not come to pass. Nor was there an occasion left for me to compare notes with Rorty on the three philosophers who, if I may express it this way, were common heroes to us. I am referring to Heidegger, Wittgenstein, and Wilfrid Sellars, and it is clear that Rorty and I came to hold widely diverging views on them. For the author of *Philosophy and the Mirror of Nature*, Heidegger was, first and foremost, a foe of "the notion of knowledge as accurate representation",[37] a philosopher whose concern was "to explore the way in which the West became obsessed with the notion of our primary relation to objects as analogous to visual perception".[38] My impression is that this dimension in Heidegger's thought never lost its primary significance for Rorty. By contrast, I came to regard the Heidegger of the 1920s as someone who genuinely has something fundamental to say about our encounter with the world, and, not incidentally, about our encounter with the visual world. It is in his *Kant and the Problem of Metaphysics* that Heidegger

---

34 Richard Rorty, *Philosophy and the Mirror of Nature*, Oxford: Basil Blackwell, 1979, *passim*, but see esp. pp. 11, 39 and 371.
35 *Ibid.*, p. 41. The expectation that "the traditional 'spectatorial' account of knowledge" might soon be "overthrown" is already voiced in Richard Rorty (ed.), *The Linguistic Turn: Recent Essays in Philosophical Method*, Chicago: The University of Chicago Press, 1967, see Rorty's "Introduction", p. 39.
36 See W. J. T. Mitchell, *Picture Theory*, Chicago: The University of Chicago Press, 1994, p. 11.
37 *Philosophy and the Mirror of Nature*, p. 6.
38 *Ibid.*, pp. 162 f.

faces the problem of how to reconcile the conceptual with the perceptual. The "power of imagination" – the Kantian *Einbildungskraft* – "refers to all representing in the broadest sense which is not in accordance with perception: conceiving of something, ... devising, having an inspiration".[39] As Heidegger puts it, "the correct understanding of the sensible character of the power of imagination" must go hand in hand with an "insight into the primary representational character of thinking".[40] Heidegger not only emphasizes that the power of imagination is a faculty which actually provides *images*,[41] but offers, in a nutshell, a brilliant analysis of the fundamental questions of pictorial representation: of what *likeness* is, and how *general images* are possible.[42]

Heidegger's book on Kant, and especially the passages I refer to here, have never been in the limelight. It is understandable that Rorty did not form a picture of Heidegger the philosopher of images. It is similarly understandable that he was unaware of the later Wittgenstein's preoccupation with pictorial representation. As Rorty put it in *Philosophy and the Mirror of Nature*: "you can't recognize a picture of X *as* a picture of X without being familiar with the relevant pictorial conventions".[43] In the heyday of linguistic philosophy, the later Wittgenstein was invariably read through Goodman's eyes. The uncontested view was that images do not depict, do not resemble; they *denote* – just like the words of verbal language. And *what* they denote will be determined by rules we have to learn.

---

39  Martin Heidegger, *Kant and the Problem of Metaphysics* (1929), transl. by Richard Taft, Bloomington: Indiana University Press, 1997, p. 91.
40  *Ibid.*, p. 103.
41  *Ibid.*, p. 91.
42  The following lines can perhaps convey the flavour of Heidegger's analyses here: "It is possible to produce a copy (photograph) ... from ... a likeness, [a photograph] of a death mask for example. The copy can now directly copy the likeness and thus reveal the 'image' (the immediate look) of the deceased himself. The photograph of the death mask, as copy of a likeness, is itself an image – but this is only because it gives the 'image' of the dead person, shows how the dead person appears, or rather how it appeared. ... – Now the photograph, however, can also show how something like a death mask appears in general. In turn, the death mask can show in general how something like the face of a dead human being appears. But an individual corpse itself can also show this. And similarly, the mask itself can also show how a death mask in general appears, just as a photograph shows not only how what is photographed, but also how a photograph in general, appears", *ibid.*, p. 66. I have corrected a misprint or mistranslation in the edition here quoted: the phrase "The copy can now directly copy the likeness" there has "only" instead of "now" (i.e. "nur" instead of "nun").
43  *Philosophy and the Mirror of Nature*, p. 25.

Now this is not at all a view Wittgenstein uniformly entertained. For instance, in the so-called "Part II" of the *Philosophical Investigations*, he outlines cases where understanding a picture appears to be entirely independent of language use. Giving the example of a "picture-face", he remarks: "In some respects I stand towards it as I do towards a human face. I can study its expression, can react to it as to the expression of the human face. A child can talk to picture-men or picture-animals, can treat them as it treats dolls." Let me note that remarks such as this were definitely rare in Wittgenstein's printed works, as available from the 1950s to the 1990s. The printed corpus only partially conveyed the richness, complexities, continuities of, and changes in, Wittgenstein's ideas on pictorial representation. It was only with the publication of the Bergen electronic edition, making his full *Nachlaß* available, that the extent of Wittgenstein's commitment to the idea of images and words playing intertwining roles became clear.[44]

# Images of Sellars

In the "Introduction" to his volume *The Linguistic Turn*, Rorty outlines a number of alternatives for the future of philosophy. One of these he characterizes as no longer envisaging "the dissolution of philosophical problems, but rather the creation of new, interesting and fruitful ways of thinking about things in general". On this alternative, "[p]hilosophers would be, as they have traditionally been supposed to be, men who gave one a *Weltanschauung* – in Sellars' phrase, a way of 'understanding how things in the broadest possible sense of the term hang together in the broadest possible sense of the term'."[45] The passage Rorty here quotes, from Sellars' "Philosophy and the Scientific Image of Man", played, way back in the late 1960s, a formative role in the development of my own thinking.[46] Sellars was my first, and most important, mentor in philosophy. We never met in person – in those

---

44  See my "Wittgenstein's Philosophy of Pictures" (2001), in A. Pichler and S. Säätelä (eds.), *Wittgenstein: The Philosopher and his Works*, Frankfurt a. M.: ontos verlag, 2006, pp. 322–353.
45  *The Linguistic Turn*, p. 34.
46  The passage in full: "The aim of philosophy, abstractly formulated, is to understand how things in the broadest possible sense of the term hang together in the broadest possible sense of the term. Under 'things in the broadest possible sense' I include such radically different items as not only 'cabbages and kings', but numbers and duties, possibilities and finger snaps, aesthetic experience and death." (Wilfrid Sellars, *Science, Perception and Reality*, London: Routledge & Kegan Paul, 1963, p. 1.)

days Hungarians were seldom permitted to leave the country for a scholarly visit to the States – but we corresponded, and he lavishly furnished me with preprints and offprints. What I was most impressed by was the particular variety of scientific realism Sellars stood for: the view that science is "continuous with common sense", and the idea of theoretical entities as postulated but real. I am still fascinated by this idea. In my rudimentary attempts to come to grips with some issues in the philosophy of time, I find Sellars' suggestion that "time has the status of a quasi-theoretical entity"[47] particularly helpful.[48] Now Sellars the scientific realist stresses that it is of course physics, or rather the future advance of physics, and not metaphysics, that ultimately determines what the nature of the theoretical entity *time* is.[49] In the final section of the present chapter I will explain why I think that on this point I have to diverge from Sellars – why I believe that we need something like descriptive metaphysics here to defend the rights of common-sense understanding in the face of apparent excesses in scientific theory-building. As I indicated above, my argument will turn on the role of *images* in our thinking.

Sellars does not allow for such a role. What he tells us in "Philosophy and the Scientific Image of Man" is that "all attempts to construe thoughts as complex patterns of images have failed, and, as we know, were bound to fail"[50], that "association of thoughts is not association of images"[51], and that "however intimately conceptual thinking is related to sensations and images, it cannot be equated with them, nor with complexes consisting of them"[52]. But Sellars does not only not *equate* thoughts with images, he actually *excludes* the latter from the realm of the former. As it becomes clear e.g. from his major essay "Empiricism and the Philosophy of Mind", mental episodes, for him, are *linguistic* episodes, and imagery boils down to *verbal* imagery[53] – while at same time, in that

---

47   Wilfrid Sellars, "Time and the World Order", in Herbert Feigl and Grover Maxwell (eds.), *Minnesota Studies in the Philosophy of Science*, vol. III, Minneapolis: University of Minnesota Press, p. 551.
48   For a first experiment of mine along Sellarsian lines see my "Time and Communication", in F. Stadler and M. Stöltzner (eds.), *Time and History/Zeit und Geschichte*, Frankfurt/M.: ontos verlag, 2006, pp. 302 f.
49   "Time and the World Order", p. 593.
50   *Science, Perception and Reality*, p. 15.
51   *Ibid.*, p. 16.
52   *Ibid.*, p. 32. I find the way this last passage begins telling: "one scarcely needs to point out these days that however intimately conceptual thinking is related to sensations and images, it cannot be equated with them…".
53   First published in 1956, repr. in *Science, Perception and Reality*, pp. 177 f.

very essay, he develops a theory within the framework of which he could easily have explained the status of mental images. According to this theory, thoughts are theoretical entities construed, in primordial times, on the analogy of overt verbal episodes. Sellars does find a place in his framework for *impressions* – but not for images.[54] Experts on Sellars might respond by pointing out that, still, the notion of "picturing" played a central role in his paper "Truth and 'Correspondence'", or indeed in the chapter on "Picturing" in his book *Science and Metaphysics*. Recall however, that for Sellars picturing was but a relation between configurations of objects in the world on the one hand, and *linguistic* configurations on the other.[55] From his reminiscences of Wittgenstein's *Tractatus* the message of paragraphs 4.016 and 4.02 is completely missing. And this is what Wittgenstein wrote there: "In order to understand the essence of the proposition, consider hieroglyphic writing, which pictures the facts it describes. – And from it came the alphabet without the essence of representation being lost. – This we see from the fact that we understand the sense of the propositional sign, without having had it explained to us."

All this is striking, for Sellars definitely had a sense for images and pictures. As becomes clear when looking at the posthumous volume *Kant and Pre-Kantian Themes: Lectures by Wilfrid Sellars*,[56] in class he loved to draw pictures and diagrams as a means to explain philosophical problems. And the situation becomes really baffling when we realize that in the mid-1930s, when Sellars was studying

---

54  A cognitive psychological theory along what can be regarded as Sellarsian lines was developed in Allan Paivio's *Imagery and Verbal Processes* (New York: Holt, Rinehart and Winston, 1971), one the first contributions to the so-called imagery debate. "Mental images", wrote Paivio, belong to the order of "postulated processes", they are "theoretical constructs", "inferential concepts", i.e. entities or processes themselves not observable, but having observable aspects and implications. Introspective experiencing of visual images on the one hand, and the objective recording of neural phenomena on the other, are empirical observations of a very different sort, but they refer to one and the same theoretical construct of a "mental image". Paivio contrasts his own methodology with "the classical approach to imagery" in which "the term image was used to refer to consciously-experienced mental processes" (*Imagery and Verbal Processes*, pp. 6–11). I will come back to Paivio later in the present chapter.

55  The title of Joseph C. Pitt's book *Pictures, Images and Conceptual Change: An Analysis of Wilfrid Sellars' Philosophy of Science* amounts to a practical joke: in this book a "picture" is defined as a "linguistic item intimately tied to the concepts of a matter-of-fact and truth" (Dordrecht: Reidel, 1981, p. 10).

56  Atascadero, CA: Ridgeview, 2002, ed. by Pedro Amaral.

philosophy at Oxford, his tutor was H. H. Price,[57] whose 1953 book *Thinking and Experience* is without doubt the fundamental twentieth-century philosophical treatise on the role of mental images. By way of ending the present section of this chapter, let me quote a passage from that book. "After listening to a lecture on Imageless Thinking", recounts Price, "a lady in the audience came up to the lecturer and said with a puzzled air, 'But, Professor, you can *think*, can't you?' ".[58]

## Believe What You Can Visualize

A famous figure that no-one assumes could not think is Albert Einstein. Now Einstein was a thoroughly visual thinker. The reader is of course familiar with those oft-quoted passages, in the Schilpp volume and in the Hadamard book, in which he insisted that in his creative work the role of the perceptual was paramount, while that of the verbal was merely secondary. "The words or the language", Einstein told Hadamard, "as they are written or spoken, do not seem to play any role in my mechanism of thought. The psychical entities which seem to serve as elements in thought are certain signs and more or less clear images… – … [These] … elements are … of visual and some of muscular type. Conventional words or other signs have to be sought for laboriously only in a secondary stage…".[59] Or the passage from his autobiographical notes: "When … memory-pictures emerge, this is not yet 'thinking'. And when such pictures form series, each member of which calls forth another, this too is not yet 'thinking'. When, however, a certain picture turns up in many such series, then … it becomes an ordering element for such series, in that it connects series which in themselves are unconnected. Such an element becomes an instrument, a concept."[60] We can assume that the visual thought-experiments through which Einstein used to explain his special theory of relativity represented pretty much the very train of thoughts that, in the first place, led him to his discoveries.

---

57  See Sellars' "Autobiographical Reflections", in Hector-Neri Castaneda (ed.), *Action, Knowledge, and Reality: Critical Studies in Honor of Wilfrid Sellars*, Indianapolis: Bobbs-Merrill, 1975, p. 285. Sellars refers to Price in "Empiricism and the Philosophy of Mind", *loc. cit.*, p. 162.
58  H. H. Price, *Thinking and Experience*, London: Hutchinson's Universal Library, 1953, p. 234. – For a brief summary of Price's position on images, see my "Pictorial Meaning and Mobile Communication", in Kristóf Nyíri (ed.), *Mobile Communication: Essays on Cognition and Community*, Vienna: Passagen Verlag, 2003, pp. 159 f.
59  Hadamard, *op. cit.* (cf. note 19 above), pp. 142 f.
60  P. A. Schilpp (ed.), *Albert Einstein: Philosopher-Scientist*, Evanston, IL: The Library of Living Philosophers, Inc., 1949, p. 7.

# Time Reduced to Space?

Based upon what he imagined – what he *visualized* – Einstein developed a view of what time really *was*. I do not take the side of Arthur Fine, who in his important essay "The Natural Ontological Attitude" (an essay Rorty seems to have found congenial,[61] and one that refers to Paul Horwich's "semantic realism" as the closest counterpart, in the philosophy of language, to Fine's own position[62]) ascribes an instrumentalist position to Einstein's 1905 paper;[63] I concur, rather, with Thomas Sattig's position, according to which "Einstein's original formulation of Special Relativity", as contrasted with the formulation he adopted under the influence of Minkowski, "was metaphysically a theory of ordinary space and time"[64]. Now the view Sattig himself accepts is the one Minkowski had put forward in 1908. As Sattig maintains: "Spacetime points and regions are not just mathematical metaphors; they are among the most fundamental entries in our ontological inventory. The realistic interpretation was adopted by Minkowski … as well as [after 1908] by Einstein".[65]

My train of thought here is designed to indicate a line of argument which might cast doubt on the reality of Minkowskian spacetime. Thus, at this juncture I shall part ways with Sattig, and join up with Arthur Fine, according to whom "to claim genuine reality for … the four-dimensional space-time manifold" amounts to accepting ideas which "not only … boggle the mind of the average man in the street …, they boggle most contemporary scientific minds as well". As Fine sees the matter, "the majority opinion among working, knowledgeable scientists" is that relativity theory is "a powerful instrument", but is not understood as a genuine foundation for "realist existence and nonexistence claims".[66]

---

61   See Rorty, "Pragmatism, Davidson and Truth" (1986), in his *Objectivism, Relativism, and Truth*, Cambridge: Cambridge University Press, 1991.

62   Fine, "The Natural Ontological Attitude" (1984), repr. in Martin Curd and J. A. Cover (eds.), *Philosophy of Science: The Central Issues*, New York: Norton, 1998, p. 1208.

63   *Ibid.*, p. 1194.

64   Thomas Sattig, *The Language and Reality of Time*, Oxford: Clarendon Press, 2006, p. 44.

65   *Ibid.*

66   Fine, *op. cit.*, pp. 1194 f. – Marshalling arguments both from the philosophy of science and the philosophy of religion, William Lane Craig, some fifteen years later, takes a similar position: "A good many philosophers of science think of the four-dimensional, geometrical representation of space-time, not realistically, but *instrumentally,* that is

Now why do I believe that the notion of a four-dimensional spacetime must, the great array of brilliant philosophical treatises to the contrary notwithstanding, indeed boggle the mind? I am coming to my main argument. In "Philosophy and the Scientific Image of Man", Sellars wrote: "it is a familiar fact that not everything that can be conceived can, in the ordinary sense, be imagined".[67] The position I am here defending, representing the tradition, in part outlined earlier in this chapter, from Plato through Hume to Titchener, Bartlett, Arnheim, H. H. Price, and Allan Paivio, maintains that, on the contrary, nothing can be conceived that cannot in the ordinary, albeit very broad, sense be imagined. Or, to put it in slightly less radical terms: scientific propositions which offer no kind of transition to visual imagery should not be taken as descriptions or indeed explanations of what there really is. By "transition", I mean something Wittgenstein seems to have meant with *überführen*, when in § 449 of *Philosophical Investigations* he wrote: "Man bedenkt nicht, daß man mit den Worten *rechnet*, operiert, sie mit der Zeit in dies oder jenes Bild überführt", or as Anscombe has it: "We do not realize that we *calculate*, operate, with words, and in the course of time translate them sometimes into one picture, sometimes into another."

In my view, the definitive theory on how words and images hang together is Paivio's *dual coding approach*, first summarized in his 1971 book *Imagery and Verbal Processes*. Paivio notes that while "the developmental studies inspired by Piaget, Bruner, and Werner all involved the assumption that images are specialized for the representation of concrete objects and events, whereas inner speech is functionally useful in dealing with abstract problems, concepts, and relationships", this functional distinction cannot be rigidly maintained, as is indicated by "the apparent development of relatively abstract (schematic) images and concretization of abstract ideas in the form of specific images". What Paivio emphasizes is that, ordinarily, "neither images nor words act as independent processes"; rather, they continually interact.[68] Now if this is what really happens, as I believe it is, then we can conclude that strings of words that do not give rise to a steady flow of images do not, strictly speaking, refer to anything; they might be symbolic devices facilitating inferences, but they in no way mirror the world.

---

to say, as an elegant and handy way of presenting the Special Theory of Relativity or the General Theory of Relativity..." (*Time and Eternity: Exploring God's Relationship to Time*, Wheaton, IL: Crossway Books, 2001, p. 95). Arguing against the notion of a timeless God, Brentano, too, consistently held that the idea that "time is the fourth dimension of space" was, at best, a *harmless fiction*. (Franz Brentano, *Philosophical Investigations on Space, Time and the Continuum*, transl. by Barry Smith, London: Croom Helm, 1988, pp. 94 ff. and 173 ff., dictations from 1915 and 1917.)

67   *Loc. cit.*, p. 5.
68   *Imagery and Verbal Processes*, pp. 27 and 32.

# The Visual and the Motor

There is an essential connection between the visual on the one hand, and the motor and the tactile on the other. Paivio reports previous research showing that mature imagery incorporates "*the implicit motor components of imitative acts*", and goes on to show that "a motor component (implicit or explicit) appears to be generally characteristic of images of movement, and of the transformations involved in the generation of an integrated figural image or the solution of more complex problems requiring visual thinking. The motor component somehow facilitates the transition from one substantive part of the stream of thought to another."[69] The two classic studies to have suggested the dependence of visual imagery on an underlying motor dimension are, first, Ribot's *Les Maladies de la volonté*, published in 1882, and secondly, the work of Galton I have referred to earlier. As Ribot has put it, "the anatomical basis of all our mental states includes both motor and sensory elements. ... our perceptions, in particular the important ones, those of sight and touch, imply as integral elements movements of the eye or the members; and ... if movement is an essential element when we see an object really, it must play the same role when we see it ideally. Images and ideas, even abstract, suppose an anatomical substratum in which the movements are in some measure represented."[70] As to Galton, he was struck by the problem that those, too, can – obviously but strangely – get along with the task of thinking who appear to be unable to experience mental images. And this was his solution to the problem: "the missing faculty seems to be replaced so serviceably by other modes of conception, chiefly, I believe, connected with the incipient motor sense, not of the eyeballs only but of the muscles generally, that men who declare themselves entirely deficient in the power of seeing mental pictures can nevertheless give life-like descriptions of what they have seen and can otherwise express themselves as if they were gifted with a vivid visual imagination".[71] It is very much under the influence of the Ribot passage just quoted that Binet introduces his discussion of the "motor type";[72] James refers to Galton, and cites Binet citing Ribot, in one single, highly important passage.[73]

---

69   *Ibid.*, pp. 30 f.
70   Théodule Armand Ribot, *The Diseases of the Will* (1882), 4th Engl. transl., Chicago: Open Court, 1915, p. 3.
71   Galton, *Inquiries into Human Faculty*, 1883 (see note 22 above), p. 61.
72   *The Psychology of Reasoning*, 1886 (see note 23 above), Engl. transl. 1899, pp. 23 f.
73   James, *op. cit.*, vol. II, p. 61.

*The motor component* – one is of course reminded here of Arnheim's analysis of *descriptive gestures*, "those forerunners of line drawing", in his 1969 book *Visual Thinking*. As he there puts it: "the perceptual qualities of shape and motion are present in the very acts of thinking depicted by the gestures and are in fact the medium in which the thinking itself takes place. These perceptual qualities are not necessarily visual or only visual. In gestures, the kinesthetic experiences of pushing, pulling, advancing, obstructing, are likely to play an important part."[74] One is reminded, also, of John M. Kennedy's 1993 book *Drawing and the Blind*,[75] providing, in fact, an elaborate new theory of visual and tactile perception. One is reminded of neurologist Antonio Damasio's remark that "[w]hen people visualize what they intend to accomplish, an accompanying bodily response makes them feel the reality of their goal".[76] And one is reminded of Hacking's insistence, in his "Experimentation and Scientific Realism", that it is not so much observability, but rather the possibility of *manipulating* objects, which is the guarantee of reality.[77] The lesson I draw from Hacking's paper is that imaginability and tangibility are closely related, and that, hence, imaginability is a likely criterion of explanatory power.

## Minkowski and Weyl

But let me return to Minkowski. This is how he began his famous talk in Cologne in 1908: "The views of space and time which I wish to lay before you have sprung from the soil of experimental physics, and therein lies their strength. They are radical. Henceforth space by itself, and time by itself, are doomed to fade away into

---

74  Rudolf Arnheim, *Visual Thinking*, Berkeley: University of California Press, 1969, pp. 117 f. I will come back to this passage by Arnheim in chapter 5 below. – The changes in Arnheim's views on the connections between the visual and the motor, occurring in the course of his journey from the first edition of his *Art and Visual Perception*, 1954, through his *Visual Thinking*, 1969, to the second edition of *Art and Visual Perception*, 1974, are as instructive as they seem to have gone practically unnoticed. I am indebted to my good friend Gábor Palló, historian of science, for memorable discussions on the subject.
75  New Haven, CT: Yale University Press.
76  Antonio R. Damasio, *Descartes' Error: Emotion, Reason, and the Human Brain*, New York: Grosset/Putnam, 1994, here quoted after Stefan Klein, *The Secret Pulse of Time: Making Sense of Life's Scarcest Commodity*, New York: Marlowe & Co., 2007, pp. 222 f.
77  Ian Hacking, "Experimentation and Scientific Realism" (1982), repr. in Curd and Cover (eds.), *Philosophy of Science*, p. 1157.

mere shadows, and only a kind of union of the two will preserve an independent reality."[78] A mere few lines later there comes the step which, from my present perspective, I see as crucial. Minkowski announces that he "will try to visualize the state of things by the graphic method". He embarks on drawing a diagram (three more will follow in the course of his presentation), saying: "With this most valiant piece of chalk I might project upon the blackboard four world-axes." And he immediately adds that understanding the diagram of course requires some abstraction, because of "the number four"; but such a measure of abstraction "is for the mathematician no infliction". By drawing this diagram, he continues, "we obtain, as an image, so to speak, of the everlasting career of the substantial point, a curve in the world, a *world-line*...".[79] The German wording is: "Wir erhalten alsdann als Bild sozusagen für den ewigen Lebenslauf des substantiellen Punktes eine Kurve in der Welt, eine *Weltlinie*...". Now the twist of course is that the "Bild" we get is not an image at all, since – forgive me for spelling out the obvious – a four-dimensional diagram cannot be drawn, cannot be visualized, cannot be imagined.[80] I guess this has been pointed out innumerable times, but let me here just refer to the 1965 paper by Peter Geach, "Some Problems about Time", observing some of the oddities of Minkowski's graphs, and let me quote from Strawson's editorial introduction to the volume in which the Geach paper was included. "Geach edges his common sense with logic", writes Strawson, "to attack some fanciful theorizing – claiming to derive respectability from physics – which, in place of our ordinary conception

---

78   Hermann Minkowski, "Space and Time", in H. A. Lorentz, A. Einstein, H. Minkowski and H. Weyl, *The Principle of Relativity: A Collection of Original Memoirs on the Special and General Theory of Relativity* (1923), Mineola, NY: Dover Publications, 1952, p. 75.
79   *Ibid.*, p. 76.
80   Clearly there exist methods of n-dimensional visualization in mathematics, a spectacular one – and probably the best known – being the parallel coordinates system by Alfred Inselberg. But Inselberg never suggested that his visualizations are as it were depictions of anything in the real world. In a telling introductory passage of his recent book *Visual Multidimensional Geometry and Its Applications*, he referred to the 1917 paper "In What Way Does it Become Manifest in the Fundamental Laws of Physics that Space has Three Dimensions?" by the physicist Paul Ehrenfest – a close friend, incidentally, of Einstein – as showing, say, that "planetary orbits are stable only in space of dimension 3. Higher-dimensional planetary systems, if they ever existed, would have a short career due to the orbits' instability, which offers an interesting hypothesis for the dimensionality of our habitat" (Alfred Inselberg, *Parallel Coordinates: Visual Multidimensional Geometry and Its Applications*, Dordrecht: Springer, 2009, p. 2).

of objects undergoing change, advocates thinking of a three-dimensional-object-at-a-time as a 'temporal slice' of a four-dimensional object. He presses his criticisms by urging the lack of analogy, the radical differences, between spatial and temporal order."[81] The Geach–Strawson pair has elicited some angry comments from J. J. C. Smart, who in a paper in 1972 stressed that although "in popular exposition" Minkowski did in fact attempt graphic visualization, "his *argument* is not the analogy with graphs. His argument is that only space-time entities are invariant…".[82] But this is precisely the point. Minkowski devised but a mathematical instrument, presenting it, however, as a true description of the real world. As Arnheim has put it in his *Visual Thinking*: while grasping the view of time suggested by the special theory of relativity can be supported by visualizing the alternation of the images of two systems, "one for which an object is in motion and another for which the same object is at rest", the "fourth spatial dimension", postulated subsequently, is "a purely mathematical construct", not accessible to our mental imagery.[83]

I will come back to Minkowski and to Smart in a minute, but let me just pause to present two famous passages by another great German mathematician, heir to the Einstein–Minkowski tradition: Hermann Weyl. The first passage is from Weyl's book *Space–Time–Matter*, originally published in 1918. "[T]he scene of action of reality", Weyl writes,

---

81  P. F. Strawson (ed.), *Studies in the Philosophy of Thought and Action*, London: Oxford University Press, 1968, p. 5.
82  J. J. C. Smart, "Space-Time and Individuals", in Richard Rudner and Israel Scheffler (eds.), *Logic & Art: Essays in Honor of Nelson Goodman*, Indianapolis: Bobbs-Merrill, 1972, p. 7. My impression is that Smart here has moved away from the position in his *Philosophy and Scientific Realism*. He there wrote: "many of the puzzles and paradoxes of relativity … can most easily be resolved by drawing diagrams of Minkowski space-time, in which most of [the] at first sight counter-intuitive facts will at once look quite obvious" (London: Routledge & Kegan Paul, 1963, pp. 136 f.).
83  Arnheim, *op. cit.*, pp. 288–291. "If a fourth spatial dimension cannot be visualized", Arnheim goes on to write, "it is probably because … [b]eyond [the third dimension] geometrical calculations – just as any other multidimensional calculations, such as factor analysis in psychology – must be content with fragmentary visualization, if any. This also means probably putting up with pieces of understanding rather than obtaining a true grasp of the whole. – No fourth dimension of space, however, is in fact claimed to exist by modern physics. It is, in the words of Arthur Eddington, 'a fictitious construction'" (*ibid.*, p. 292). The piece by Eddington Arnheim refers to is the chapter "Spherical Space" in the former's *The Expanding Universe* (1933), Arnheim quotes from the collection by Milton K. Munitz (ed.), *Theories of the Universe: From Babylonian Myth to Modern Science*, Glencoe, IL: The Free Press, 1957.

is not a three-dimensional Euclidean space, but rather a *four-dimensional world, in which space and time are linked together indissolubly.* However deep the chasm may be that separates the intuitive nature of space from that of time in our experience, nothing of this qualitative difference enters into the objective world which physics endeavours to crystallise out of direct experience. It is a four-dimensional continuum, which is neither "time" nor "space". Only the consciousness that passes on in one portion of this world experiences the detached piece which comes to meet it and passes behind it, as *history*, that is, as a process that is going forward in time and takes place in space.[84]

The second passage is from Weyl's *Philosophy of Mathematics and Natural Science*, originally published in 1927 in German. As Weyl here puts it: "The objective world simply *is*, it does not *happen*. Only to the gaze of my consciousness, crawling upward along the life line of my body, does a section of this world come to life as a fleeting image in space which continuously changes in time."[85] What Richard Gale says about this passage does just as well fit the first one, namely that it should be understood as a metaphor, since, if taken literally, it would be simply absurd.[86] But let me make two comments. First, that whether taken literally or not, these passages are metaphysical statements, not implied by the mathematics on which they are apparently based. This is especially conspicuous in the case of the 1927 formulation, following in the book after an extended, partisan philosophical argument. My second comment is that metaphors are meaningless if they cannot be visualized, as Weyl's obviously cannot. I conclude that the Minkowski–Weyl interpretation of space-time is a merely instrumental one. And I suggest that scientific explanations must end, and a common-sense world view ought to be defended, at the point where mathematics ceases to be backed by images.

## Visualization in Mathematics

Clearly, most of mathematics *is* backed by images. This is – may I refer back to the beginning of the present chapter – rather evident in the case of geometry. Arnheim himself, in his chapter "Thinking with Pure Shapes", stressed that not only "self-evident geometry", but also arithmetics and algebra have a thoroughly

---

84 Hermann Weyl, *Space–Time–Matter* (4th German edition 1921), Mineola, NY: Dover Publications, 1952, p. 217.
85 Hermann Weyl, *Philosophy of Mathematics and Natural Science*, Princeton, NJ: Princeton University Press, 1949, p. 116.
86 Richard M. Gale (ed.), *The Philosophy of Time: A Collection of Essays* (1967), London: Macmillan, 1968, pp. 298 f.

perceptual basis, that "[c]ounting is preceded by the perceptual grasp of groups", and that "[n]umbers are perceptual entities, visual and to some extent tactual and auditory".[87] Recent developments suggest that the 19-century visualization Angst in mathematics, and in the philosophy of mathematics, is receding. In his book *Visual Thinking in Mathematics*, Marcus Giaquinto convincingly argues that it is indeed possible "to achieve generality when thinking with particular images"[88] – a geometrical proof *can*, and when possible, *should*, proceed visually; that "[s]o far from being language based, the origin of our knowledge of simple sums seems to be a kind of finger expertise"[89], and both arithmetics and number theory allow for visual proofs;[90] that in algebra "[s]ubstitution, relocation, copying, deletion, and insertion" – that is, the "major classes of symbol manipulation" – are typically "performed in visual imagination, when moving from one term or formula to another. It is likely that in *some* cases, especially symbol relocation, the visualizing has a motor element";[91] and that even in analysis there is room and need for visualization – Giaquinto refers to, and elucidates, the famous Cambridge mathematician J. E. Littlewood's piece "Post-script on pictures". Littlewood, Giaquinto writes, did indeed believe that "a diagram could provide proof of an analytic theorem".[92]

---

87  Arnheim, *op. cit.*, pp. 221 f., 211 and 213. On p. 214 Arnheim refers to Marguerite Lehr's highly interesting introduction to Catherine Stern's seminal book, *Children Discover Arithmetic: An Introduction to Structural Arithmetic*, London: George G. Harrap, 1953. Catherine Stern was, at the New School for Social Research from 1940 to 1943, research assistant to Max Wertheimer, the founder of Gestalt Psychology.
88  Marcus Giaquinto, *Visual Thinking in Mathematics: An Epistemological Study*, Oxford: Oxford University Press, 2007, p. 151.
89  *Ibid.*, p. 123.
90  On this issue see also Michael D. Resnik, *Mathematics as a Science of Patterns*, Oxford: Clarendon Press, 1997, cf. esp. pp. 229 ff.
91  Giaquinto, *op. cit.*, p. 203.
92  *Ibid.*, p. 163. – As Littlewood puts it in the section "Post-script to pictures", in the volume *Littlewood's Miscellany* (Cambridge: Cambridge University Press, 1953, repr. 1986, p. 54): "My pupils *will* not use pictures, even unofficially and when there is no question of expense. This practice is increasing; I have lately discovered that it has existed for 30 years or more, and also why. A heavy warning used to be given that pictures are not rigorous; this has never had its bluff called and has permanently frightened its victims into playing for safety. Some pictures, of course, are not rigorous, but I should say most are (and I use them whenever possible myself)." Littlewood and Wittgenstein were friends. They first met in Manchester, and then again in Cambridge. The greater part of Wittgenstein's numerous drawings in his manuscripts pertain to the foundations of mathematics; and a major message of Wittgenstein's

The point where visualization in mathematics utterly breaks down is where it purports to picture time as a fourth dimension of space. It is here the common-sense world-view has to step in. To defend the common-sense world-view involves explaining, without explaining away, some crucial common-sense metaphors. Now this is how J. J. C. Smart begins his 1949 paper "The River of Time":

> There are certain metaphors which we commonly feel constrained to use when talking about time. We say that we are advancing through time, from the past into the future, much as a ship advances through the sea into unknown waters. Sometimes, again, we think of ourselves as stationary, watching time go by, just as we may stand on a bridge and watch leaves and sticks float down the stream underneath us. ... Thus instead of speaking of our advance through time we often speak of the flow of time. ... These metaphorical ways of talking are philosophically important in a way in which most metaphorical locutions are not. They ... are, in some way, *natural* to us; at first sight, at any rate, it seems difficult to see how we could avoid them.[93]

Difficult or not, Smart did his best to demonstrate the alleged spuriousness of these common-sense metaphors. By contrast, I believe we should strive to build up a philosophical strategy which in fact vindicates them. Such a strategy is hinted at in chapter 4 below. Coming to the end of the present chapter, there remain three questions.

## Concluding Queries

The first question, lurking in the background throughout my argument: what does "imaginability" amount to? Is imaginability confined to what we, in our world as it is actually given to us, can in fact imagine? Should we not, rather, say what Reichenbach, referring to Helmholtz, suggests, namely that "imagining ... visually" a world different from ours is indeed possible, by "depicting the series of sense perceptions which one would have if one lived in such a world"; and that "human beings, living in a non-Euclidean world, would develop an ability of visualization which would make them regard the laws of non-Euclidean geometry as necessary and self-evident, in the same fashion as the laws of Euclidean geometry appear self-evident to us."[94] My stance here is similar to Ramsey's, commenting

---

    philosophy of mathematics is that mathematical facts and physical facts *overlap*; images of physical facts, then, can indeed convey mathematical truths.

93    J. J. C. Smart, "The River of Time", *Mind*, vol. 58, no. 232 (Oct. 1949), p. 483.
94    Hans Reichenbach, "The Philosophical Significance of the Theory of Relativity", in Schilpp (ed.), *Albert Einstein*, pp. 300 and 308.

on the *Tractatus*: "what we can't say we can't say, and we can't whistle it either". What we can't imagine we can't imagine, and can't whistle it either. We can imagine, for we can visualize, spherical geometry, although it constitutes a kind of non-Euclidean one. But we can in no way visualize, say, spacetime with eleven dimensions as string theory suggests; and here we should not let physicists string us along, but should assume a decidedly instrumentalist attitude.

Secondly, with the views I have here put forward, where would I locate my position in the philosophy of science? I still feel myself belonging to the realist camp, siding with Grover Maxwell's contention that there is a "continuous transition from observability to unobservability" which has no relevance at all to the existence/nonexistence issue;[95] siding with Hacking's view that the experimenter is necessarily, and rightly, convinced of the reality of a great many unobservable entities;[96] and of course sharing Sellars' faith in the power of science to draw up ever more correct images of the world. But I also take seriously the cautioning words Sellars again and again voiced: what contemporary science offers consists, in no small measure, of promissory notes. The position I suggest appears to me to offer a felicitous compromise between commonsense realism and scientific realism. But there is one variety of scientific realism I can, clearly, not make friends with: *structural realism*, although this is considered, it seems, by many realists and antirealists alike as "the most defensible form of scientific realism"[97]. Structural realism says that we should epistemically commit ourselves only to the mathematical or structural content of our theories. I believe that, quite on the contrary, we should commit ourselves to the visualizable content of them.

And so, by way of conclusion, let me now ask myself a third, very brief, and somewhat emotional, question: had that hoped-for next meeting with Rorty happened, how would it have played out? Certainly we would have been in agreement that verbal language in general, and the language of theories in particular, do *not picture*; they are conventional instruments the community of human beings use.

---

95  Grover Maxwell, "The Ontological Status of Theoretical Entities" (1962), repr. in Curd and Cover (eds.), *Philosophy of Science*, p. 1057.
96  Hacking, *op. cit.*, pp. 1154 f.
97  See the opening sentence of James Ladyman's excellent discussion in the *Stanford Encyclopedia of Philosophy* entry "Structural Realism" (http://plato.stanford.edu/entries/structural-realism, first published in Nov. 2007), see also Ladyman's earlier book *Understanding Philosophy of Science*, London: Routledge, 2002, as well as the unpublished, but widely cited, chapter "The Scientific Realism Debate" in Ioannis Votsis' 2004 PhD dissertation.

But everyday thinking and communication, as well as scientific theories, involve more than just verbal language. They involve images, too. They involve, indeed they fundamentally rely on, visualizations. And can we not say that the images presupposed, or suggested, by our most successful theories amount to something like *mirrors of nature*? I believe Rorty would have found this idea intriguing. He might even have liked it.

# 2. Hundred Years After: How McTaggart Became a Thing of the Past

In chapter 1 above I had occasion to express my indebtedness to the man who had been my first mentor in philosophy, one who played a significant role in analytic philosophy from the late 1940s to the 1960s, and one who was renowned for his skill in exploiting the history of philosophy as the background against which to act out philosophical analysis: Wilfrid Sellars. The present chapter, too, owes much to the writings of Sellars. In a passage in his "Autobiographical Reflections" Sellars describes his first serious encounter with philosophy. It happened at Ann Arbor, in 1931/32, when in a seminar in metaphysics he was introduced, as he reports, "to McTaggart's classic paper on the unreality of Time", and chose to write his term paper on the topic. He was soon "deep in the literature" and found himself "genuinely involved". As he puts it: "Philosophy was no longer a storehouse of alternatives to be explored and evaluated but, from that moment on, an unfinished dialogue in which I might have something to say. I soon became convinced that the problem of time was so intimately connected with other classical problems that it, like the mind-body problem, is one of the major proving grounds for philosophical systems."[1] Sellars continued to work on the topic of time, returning to it again and again; and defending, from the very beginning, "a substantialist ontology of change", that is, a position diametrically opposed to that of McTaggart. I will come back to Sellars on two occasions later in this chapter; just now, let me give a summary outline of the same.

McTaggart's paper on "The Unreality of Time" was published in 1908, in the journal *Mind*. The argument of the paper is sufficiently elusive to stand in need of scrutiny before being subjected to criticism. Such scrutiny is what I will attempt to provide in the first section of the chapter, under the heading "The McTaggart Motley". In the second section, under the heading "Refuted and Ridiculed", I shall summarize the devastating criticisms that, since the 1920s, C. D. Broad, and others in his wake, have been directing against McTaggart's position, asking, in the third section, how, in the face of such a series of convincing refutations, his argument

---

1 Wilfrid Sellars, "Autobiographical Reflections", in H.-N. Castaneda (ed.), *Action, Knowledge, and Reality: Critical Studies in Honor of Wilfrid Sellars*, Indianapolis: Bobbs-Merrill, 1975, p. 281.

could still gain, and does still gain, adherents. The answer is, as I will briefly show, that McTaggart's position has become mixed up with, and won undeserved respectability from, the Einstein–Minkowski conception of space-time, proclaimed in the very same year that McTaggart's paper was published. In the final section of the chapter I shall sketch, under the heading "A Future for Time?", the rudiments of an alternative – admittedly adventurous – philosophical strategy, designed to overcome the position represented by McTaggart; that is, to vindicate the common-sense view of the reality of time.

## The McTaggart Motley

McTaggart's paper exists in two versions – or in two-plus-a-bit versions, if you like. The first one is the *Mind* version.[2] The second, bearing the title "Time", is the text making up chapter XXXIII in the second volume of McTaggart's *The Nature of Existence*, published in 1927. This was a posthumous publication. McTaggart died in 1925, leaving behind a semi-finished draft of the volume, half typescript, half manuscript, bequeathing to C. D. Broad, his successor at Trinity College, Cambridge, the task of preparing it for press. Bringing it into line with the first volume that had been published in 1921, Broad divided the text into numbered sections, constructed an analytical table of contents, but otherwise reports to have made only very minor editorial changes.[3] Perhaps he should have been more thorough. Chapter XXXIII was printed from the typescript part of the draft, but my impression is that the typescript had not been without flaws, with some resulting wordings even more confused than McTaggart's formulations usually were. Also, it is generally unrecognized that the textual differences between the 1908 paper and the *Nature of Existence* version are quite significant. Certainly the latter is not just a re-written text of the former. Rochelle's formula, according to which the "Unreality of Time" paper "[f]orms a substantial part" of *The Nature of Existence* chapter, is closer to the facts.[4] For instance, the so-called "C series", the discussion of which McTaggart clearly saw as playing an important role in the overall argument of the 1908

---

2     J. Ellis McTaggart, "The Unreality of Time", *Mind: A Quarterly Review of Psychology and Philosophy*, N.S., no. 68, October 1908, pp. 457–474.
3     Cf. the "Editor's Preface", p. v, in John McTaggart Ellis McTaggart, *The Nature of Existence*, vol. II, ed. by C. D. Broad, Cambridge: Cambridge University Press, 1927.
4     Gerald Rochelle, *The Life and Philosophy of J. McT. E. McTaggart, 1966–1925*, Lewiston, NY: Edwin Mellen, 1991, p. 234.

paper, is introduced only in the last paragraphs of the 1927 "Time" chapter, the topic then recurring, with embellishments, in later chapters of the volume. In the 1927 chapter, there is an extended analysis directed against Russell's treatment of time in his 1903 book *The Principles of Mathematics*, entirely missing in the 1908 paper. More importantly, the 1927 chapter contains a five-page discussion of the criticism C. D. Broad levelled, in his 1923 book *Scientific Thought*, at McTaggart's 1908 position. To mention one more example, while in the 1908 paper the hypothesis that "there might be several independent time-series in reality" is introduced as a possibility raised by Bradley, and the implication that under such conditions "no time would be *the* time – it would only be the time of a certain aspect of the universe" is rejected with reference to the fact that "the theory of a plurality of time-series is a mere hypothesis" and "no reason has ever been given why we should believe in their existence", in the 1927 chapter the name of Bradley is missing, and the observation that under the conditions discussed "no time would be *the* time – it would only be the time of a certain aspect of the universe" is not followed by the remark that no reason has ever been given for the hypothesis in question. Why the change? Might it not be Einstein, after all, who haunts McTaggart here? Might not, by the 1920s, the news about the special theory of relativity, against all the odds, have reached him? But I am getting ahead of myself. I said McTaggart's paper exists in two-plus-a-bit versions; I managed to list the first two; I am now coming to the plus-a-bit one. This is the reprint of "The Unreality of Time" in the volume *Philosophical Studies*, a 1934 collection of McTaggart's essays.[5] I am calling it a plus-a-bit version, because although it is indeed a reprint, it is supplemented by a number of notes by the editor S. V. Keeling, indicating the places where the *Nature of Existence* text contains significant additions to the 1908 one. Even if not conveying the full extent of the differences between the first two versions, these notes are interesting. Interesting, or rather, telling, is also the chapter "The Relation of Time and Eternity" in *Philosophical Studies*, following upon the "Unreality of Time" chapter. This is the text of a talk delivered by McTaggart before the Philosophical Union of the University of California on August 23, 1907. I am tempted to call it version zero of the 1908 *Mind* paper, giving a feel, as it were, of the *weltanschauung* behind the latter. As McTaggart here put it:

> All existence which presents itself as part of our ordinary world of experience presents itself as temporal. But ... we have reason to believe that some reality which exists, exists timelessly – not merely in the sense that its existence endures through

---

5   J. McTaggart Ellis McTaggart, *Philosophical Studies*, ed., with an introduction, by S. V. Keeling, London: Edward Arnold, 1934.

unending time, but in the deeper sense that it is not in time at all. ... I do see a possibility of showing that the timeless reality would be, I do not say unmixedly good, but very good, better than anything which we can now experience or even imagine. I do see a possibility of showing that all that hides this goodness from us – in so far as it is hidden – is the illusion of time.[6]

This passage, glaringly mystical and devoid of analytic rigour, might give us a foretaste of McTaggart's arguments in "The Unreality of Time". It is an inventory of these arguments I now turn to.

I am speaking of "arguments" in the plural, since I believe that McTaggart's essay cannot be seen – contrary to what standard summaries take for granted – as proceeding along a single train of thought. It consists, rather, of a number of sometimes overlapping, sometimes frayed and only loosely connected, threads – stipulations, arguments, half-arguments, and asides. Attempting to take stock of them here, I cannot avoid repeatedly quoting McTaggart's text directly. Commenting on McTaggart's favourite formula that if an historical event is ever earlier than another, then it always was and will be earlier, Miss Cleugh in her 1937 book *Time and Its Importance in Modern Thought* says that this is "an unsatisfactory way of expressing" whatever McTaggart wishes to convey, "and one which is perilously near nonsense".[7] My impression is that McTaggart's wordings are almost always perilously near nonsense, not yielding to meaningful and yet faithful paraphrase; hence my preference for direct citations. Let me first quote the string of stipulations McTaggart begins his essay with. "Positions in time", writes McTaggart, "as time appears to us *prima facie*, are distinguished in two ways. Each position is Earlier than some, and Later than some, of the other positions. And each position is either Past, Present, or Future. The distinctions of the former class are permanent, while those of the latter are not. If M is ever earlier than N, it is always earlier. But an event, which is now present, was future and will be past." McTaggart then goes on to refer to "the series of positions running from the far past through the near past to the present, and then from the present to the near future and the far future, as the A series"; the "series of positions which runs from earlier to later" he calls "the B series"; and he concludes the passage with the stipulations "[t]he contents of a position in time are called events", and "[a] position in time is called a moment".[8] With this passage – let me list it as THE A AND B SERIES STIPULATION – the

---

6   *Ibid.*, p. 135.
7   M. F. Cleugh, *Time and Its Importance in Modern Thought*, London: Methuen, 1937, p. 153.
8   McTaggart, "The Unreality of Time", *Mind*, 1908, p. 458.

stage is set; by accepting it as a point of departure, the reader accepts an idiosyncratic – namely *timeless* – way of speaking about temporal phenomena. McTaggart now continues by pressing the point that "the A series is essential to the nature of time". As he puts it, "a B series without an A series" will not suffice to "constitute time", and, consequently, if "the distinction of past, present and future" is an illusion, then *time* must be an illusion, too. He puts forward here what might be taken as his first attempted proof of the unreality of time – I am listing it as the EVENTS NEVER CHANGE argument. This is how it runs: "It would, I suppose, be universally admitted", writes McTaggart, "that time involves change. ... A universe in which nothing whatever changed ... would be a timeless universe. – If, then, a B series without an A series can constitute time, change must be possible without an A series. Let us suppose that the distinction of past, present and future does not apply to reality. Can change apply to reality? What is it that changes?" McTaggart insists that what *cannot* change are events. "An event", as he puts it, "can never cease to be an event. ... it will always be, and has always been, an event, and cannot begin or cease to be an event." On the other hand, indicates McTaggart, events change in the sense that future events become present events, and present events become past events. I am citing an oft-quoted passage:

> Take any event – the death of Queen Anne, for example – and consider what change can take place in its characteristics. That it is a death, that it is the death of Anne Stuart, that it has such causes, that it has such effects – every characteristic of this sort never changes. ... in every respect but one it is ... devoid of change. But in one respect it does change. It began by being a future event. It became every moment an event in the nearer future. At last it was present. Then it became past, and will always remain so, though every moment it becomes further and further past.

Now this kind of change, McTaggart tells us, can only be posited if we assume there to be an "A series". No time without change, and no change without the "A series".[9]

The next step to follow is the introduction of the "C series", a series that is "not temporal, for it involves no change, but only an order".[10] McTaggart puts forward an argument that purports to show that "the A series, together with the C series, is sufficient to give us time. ... It is", he writes, "when the A series, which gives change and direction, is combined with the C series, which gives permanence, that the B series can arise."[11] I do not wish to spend time on this argument here – let me call it the A PLUS C MAKE B argument – but let me just remark, however, that

---

9   *Ibid.*, pp. 458–461.
10  *Ibid.*, p. 462.
11  *Ibid.*, pp. 463 f.

it is quite usual for commentaries not to take note of it, nor even of the "C series" as such. Alexander Gunn in his classic *The Problem of Time*[12] does not; Gregory Currie in his 1992 essay "McTaggart at the Movies"[13] does not; Runggaldier in his 2005 paper "Are There 'Tensed' Facts (A-Series)?"[14] does not; Kanzian in his 2005 paper "Warum McTaggarts Beweis für die Unwirklichkeit der Zeit fehlschlägt"[15] does not; Katalin Farkas in her *Synthese* paper "Time, Tense, Truth"[16] does not; Richard Gale in *The Blackwell Guide to Metaphysics*[17] does not. Indeed Gale in his reader *The Philosophy of Time*[18] prints McTaggart's 1927 "Time" chapter with the last pages – the pages where the "C series" are introduced – left out. McTaggart might have believed that his arguments add up to a cohesive whole, but many of his commentators clearly thought otherwise. They were right. Upon the A PLUS C MAKE B argument there follows, in the 1908 text, the digression on the possible plurality of time-series[19] I have referred to above – let me list it as THE MULTIPLE TIMES ASIDE; then comes an entirely obscure passage which I shall christen THE A SERIES ARE RELATIONS OF EVENTS half-argument, and which McTaggart concludes with the words, "[t]he relations which form the A series … must be relations of events and moments to something not itself in the time-series. What this something is might be difficult to say"[20]; and upon this half-argument then follows what might be regarded as the main argument of the essay "The Unreality of Time" – I will call it the IMPOSSIBILITY OF THE A SERIES argument.

Presenting this argument I must, again, quote McTaggart at some length. "Past, present, and future", he writes, "are incompatible determinations. Every event must be one or the other, but no event can be more than one. … If M is past, it has been present and future. If it is future, it will be present and past. If it is present, it has

---

12    J. Alexander Gunn, *The Problem of Time: An Historical and Critical Study*, New York: Richard R. Smith, 1930, pp. 345–349.
13    *Philosophy*, vol. 67, no. 261 (July 1992), pp. 343–355.
14    In F. Stadler and M. Stöltzner (eds.), *Time and History/Zeit und Geschichte*, Frankfurt/M.: ontos verlag, 2006, pp. 77–84.
15    In F. Stadler and M. Stöltzner (eds.), *Time and History: Papers of the 28th International Wittgenstein Symposium*, Kirchberg am Wechsel: ALWS, 2005, pp. 131–133.
16    *Synthese*, vol. 160, no. 2 (January 2008), pp. 269–284.
17    Cf. Richard M. Gale, "Time, Temporality, and Paradox", in R. M. Gale (ed.), *The Blackwell Guide to Metaphysics*, Oxford: Blackwell Publishers, 2002, pp. 66–86.
18    Richard M. Gale (ed.), *The Philosophy of Time: A Collection of Essays* (1967), London: Macmillan, 1968.
19    "The Unreality of Time", p. 466.
20    *Ibid.*, p. 468.

been future and will be past. Thus all the three incompatible terms are predicable of each event, which is obviously inconsistent with their being incompatible..." Now it might be objected, McTaggart says, that this is only a seeming incompatibility. An adversary might point out that "our language has verb-forms for the past, present, and future, but no form that is common to all three. It is never true, the answer will run, that M *is* present, past and future. It *is* present, *will be* past, and *has been,* future. Or it *is* past, and *has been* future and present, or again *is* future and *will be* present and past. The characteristics are only incompatible when they are simultaneous, and there is no contradiction to this in the fact that each term has all of them successively."[21] McTaggart retorts, and purports to prove in some detail, that this objection involves a vicious circle – let me, then, list the passages involved as the VICIOUS CIRCLE argument. I must admit that I am unable to follow him here; that I am glad every time I encounter a commentary refuting the VICIOUS CIRCLE argument; but that, generally speaking, I am not able to follow those refutations either. However, I think I am able to follow, and I take pleasure in, the remaining two arguments, or semi-arguments, that the "Unreality of Time" essay offers. These are, first, the SPATIAL MOVEMENT METAPHOR FOOTNOTE, and, secondly, the SPECIOUS PRESENT argument.

In the SPATIAL MOVEMENT METAPHOR FOOTNOTE, there are unmistakable echoes of Bradley. One is reminded of the *Principles of Logic* passage, "the present is no time[;] … it is a point we take within the flow of change";[22] or of the *Appearance and Reality* passages, "[i]t is usual to consider time under a spatial form. It is taken as a stream, and past and future are regarded as parts of it… It is natural to set up a point in the future towards which all events run, or from which they arrive, or which may seem to serve in some other way to give direction to the stream. … We think forward, one may say, on the same principle on which fish feed with their heads pointing up the stream."[23] This is how the SPATIAL MOVEMENT METAPHOR FOOTNOTE runs, and I am not quoting the passage in full:

> It is very usual to present Time under the metaphor of a spatial movement. But is it to be a movement from past to future, or from future to past? … If the events are taken as moving by a fixed point of presentness, the movement is from future to past, since the future events are those which have not yet passed the point, and the past are those which have. If presentness is taken as a moving point successively related to each of a series of events, the movement is from past to future. Thus we say that events come out of the future, but we say that we ourselves move towards the future. For each man

---

21  *Ibid.*
22  F. H. Bradley, *The Principles of Logic*, London: Oxford University Press, 1883, Bk. I, p. 53.
23  F. H. Bradley, *Appearance and Reality*, London: Swan Sonnenschein, 1893, pp. 39 and 214.

identifies himself especially with his present state, as against his future or his past, since the present is the only one of which he has direct experience. And thus the self, if it is pictured as moving at all, is pictured as moving with the point of presentness along the stream of events from past to future.[24]

I take the SPATIAL MOVEMENT METAPHOR FOOTNOTE to be understood by McTaggart as a third proof of the unreality of time, further supporting, as it were, the IMPOSSIBILITY OF THE A SERIES argument and the VICIOUS CIRCLE argument. If the passage of time were real, McTaggart must have thought, the direction of time's flow would be unambiguously given. The fact that time appears to us as a movement both "from past to future" and "from future to past" proves that that movement is, indeed, mere appearance. However, I might think of a second, rather more interesting, reading of the SPATIAL MOVEMENT METAPHOR FOOTNOTE. On this reading, Bradley, and subsequently McTaggart, have discovered what later, in the 1980s, became one of the important findings of conceptual metaphor theory, namely that there are two related, but apparently different, ways to conceptualize time: the "time-moving" and the "ego-moving" metaphors. As I will attempt to show in the last section of the present chapter, that finding could play a significant role in a philosophical strategy designed to demonstrate the *reality* of time. Just now, however, by way of concluding this section, let me discuss, very briefly, McTaggart's SPECIOUS PRESENT argument.

The term "specious present" was coined by E. R. Clay in 1882, and made more precise by William James in his *The Principles of Psychology*, published in 1890. As James in an oft-cited passage puts it, "the practically cognized present is no knife-edge, but a saddle-back, with a certain breadth of its own on which we sit perched, and from which we look in two directions into time. The unit of composition of our perception of time is a *duration*…"[25] To express it in a nutshell, the notion of the specious present is the empirically supported alternative to the age-old speculative notion of the present as a fleeting, momentary boundary between the future and the past. McTaggart of course cannot accept this latter notion, since he does not believe either in the future or in the past; while he does accept the experience of the specious present as an empirical fact. However, as he points out, "the 'specious present' varies in length according to circumstances, and may be different for two people at the same period. The event M may be simultaneous both with X's perception Q and Y's perception R. At a certain moment Q may have ceased to be part of X's specious present. M, therefore, will at that moment be past. But

---

24  "The Unreality of Time", p. 470.
25  William James, *The Principles of Psychology* (1890), London: Macmillan & Co., 1901, vol. I, p. 609.

at the same moment R may still be part of Y's specious present. And, therefore, M will be present, at the same moment at which it is past. This", McTaggart says, "is impossible."[26] What the phenomenon of the specious present according to McTaggart demonstrates is, precisely, that time is illusory; accepting the reality of time, he tells us again by way of conclusion, leads to paradoxical results.

## Refuted and Ridiculed

At the very beginning of his 1908 paper, McTaggart has some lines explaining that the doctrine of the unreality of time is not at all an unheard-of one; in fact "in all ages" it has been "singularly attractive" – or "singularly persistent", as he puts it in the 1927 version, in which these lines are repeated with some slight changes only. McTaggart refers to the philosophy, religion, theology and the mysticism of the East and West; mentioning, in particular, the philosophers Spinoza, Kant, Hegel, Schopenhauer, and Bradley. He could also have referred to, say, Parmenides, Zeno of Elea, Augustine, or, among the moderns, Leibniz. In fact, the view that time is somehow real has always been a minority position in philosophy,[27] defended, with reservations, by Aristotle, and postulated, rather than demonstrated, by Newton. Time was real, indeed it was the ultimate reality, for Henri Bergson, writing at the turn of the nineteenth and twentieth centuries; but Bergson had, for understandable reasons, almost no impact on analytically minded philosophers. Russell even wrote a pamphlet against him in 1914. But he did influence C. D. Broad; and William James of course adored him. Be that as it may, McTaggart might well have been unaware of Bergson in 1908, and even in later years. And he was entirely right when depicting the doctrine of the unreality of time as a mainstream one. Also, he was right in maintaining that his own arguments – or his own "reasons", as he puts it[28] – for the denial of the reality of time were different from those employed by other philosophers. But he was mistaken in believing that his arguments were sound. I am now coming to the criticism that C. D. Broad, in the 1920s and 1930s, has levelled at McTaggart.

In his "Intellectual Autobiography", Broad recalls his student days at Cambridge, roughly at the time McTaggart published his *Mind* essay. McTaggart was

---

26 "The Unreality of Time", p. 472.
27 Cf. the section "A Nutshell History of the Philosophy of Time", in my paper "Time and the Mobile Order", in Kristóf Nyíri (ed.), *Mobile Studies: Paradigms and Perspectives*, Vienna: Passagen Verlag, 2007, pp. 103–105.
28 "The Unreality of Time", p. 457.

one of the teachers "from whose lectures and personal instruction [he] gained most". However, apparently it was easier to venerate McTaggart than to build on his work. As Broad writes: "No one could fail to be impressed by his extraordinary dialectical power, his wit, and his amazing quickness in discussion; but, though he had many admirers, he had hardly any disciples. For all practical purposes Moore and Russell held the philosophical field and continued to do so for many years."[29] After teaching at St. Andrews, Dundee, and Bristol, Broad became McTaggart's successor at Trinity College in 1923. The same year, he published his book *Scientific Thought*. In this book, he takes up "the alleged difficulty that every event is past, present, and future; that these characteristics are incompatible; and that there is no way of reconciling them which does not either involve an infinite regress, in which the same difficulty recurs at every stage, or a vicious circle. This argument", Broad writes, "has been used by Dr M'Taggart as a ground for denying the reality of Time. It is certainly the best of the arguments which have been used for this purpose; since it really does turn on features which are peculiar to Time, and not, like most of the others, on difficulties about continuity and infinity which vanish with a knowledge of the relevant mathematical work on the subject."[30] May I just interject, though the issue has no direct bearing on our present topic, that Broad is here victim to a widespread error; as Whitrow in his magnificent book *The Natural Philosophy of Time* explains, Cantor did not solve Zeno's problem.[31] But back to McTaggart. Broad goes on by referring to the EVENTS NEVER CHANGE argument, citing the "example of the death of Queen Anne, as an event which is supposed to combine the incompatible characteristics of pastness, presentness, and futurity". Broad's comment is momentous. "[F]uturity", he says, "is not and never has been literally a characteristic of the event which is characterised as the death of Queen Anne. Before Anne died, there was no such event as Anne's death, and 'nothing' can have no characteristics."[32] The criticism levelled at McTaggart, as Broad here advances it, must be seen against the background of the latter's own philosophy of time and change. According to this philosophy, it of course makes sense to speak of the changes of *things*, but not of the changes of events.[33] "When an event,

---

29  In *The Philosophy of C. D. Broad*, ed. by P. A. Schilpp, New York: Tudor Publishing, 1959, p. 50.
30  C. D. Broad, *Scientific Thought*, London: Kegan Paul, Trench, Trubner & Co., 1923, p. 79.
31  G. J. Whitrow, *The Natural Philosophy of Time*, London: Thomas Nelson, 1961, pp. 135 and 145–148.
32  Broad, *Scientific Thought*, pp. 79 f.
33  *Ibid.*, pp. 62 ff.

which was present, becomes past", writes Broad, "it does not change or lose any of the relations which it had before; in simply acquires in addition new relations which it *could* not have before, because the terms to which it now has these relations were then simply non-entities. – It will be observed", Broad continues, "that such a theory as this accepts the reality of the present and the past, but holds that the future is simply nothing at all. Nothing has happened to the present by becoming past except that fresh slices of existence have been added to the total history of the world." This increase in "the sum total of existence" is what Broad calls *becoming*.[34] "[T]he laws of logic", Broad maintains, "apply to a fixed universe of discourse… But the universe of actual fact is continually increasing through the becoming of fresh events; and changes in truth, which are mere increases in the *number of truths* through this cause, are logically unobjectionable." Contrary to what McTaggart believed, Broad says, "no event ever does have the characteristic of futurity", and it is because of this that the law of the excluded middle does not apply to future events.[35]

Broad repeats these same critical observations in greater detail, and in rather harsher terms, in the second volume of his book *Examination of McTaggart's Philosophy*, published in 1938.[36] The text he there analyzes, in the chapter "Ostensible Temporality", is the 1927 version of McTaggart's paper; but his remarks fully apply to the 1908 version, too. He dwells at some length on McTaggart's attempt to replace all temporal copulas by a single non-temporal one. Referring to the EVENTS NEVER CHANGE argument, and to the McTaggartian formula that if an historical event *ever* precedes another historical event by a given interval, than it *always* precedes the latter by exactly that interval, Broad says that "[n]o one but a philosopher doing philosophy" would use the verb "precedes" in this seemingly non-temporal sense. "Such phraseology", points out Broad, "would suggest that the two events are particulars which (*a*) somehow *coexist* either timelessly or simultaneously, and yet (*b*) stand timelessly or sempiternally in a certain *temporal* relation of precedence. This must be nonsense, and it is most undesirable to use phrases which inevitably suggest such nonsense. I cannot help suspecting", writes Broad, "that there is some muddle of this kind at the back of McTaggart's mind when he says that events cannot be annihilated or generated because this would be incompatible with the fact that they *always* stand in the determinate temporal

---

34  *Ibid.*, pp. 66 f. Any "complete analysis of the qualitative changes of things", Broad here points out, "is found to involve the coming into existence of events" (*ibid.*, p. 67).
35  *Ibid.*, pp. 83 and 81.
36  Cambridge: Cambridge University Press, 1938. The first volume appeared in 1933.

relation in which they do stand to each other."³⁷ Coming to the end of the chapter "Ostensible Temporality", Broad sums up McTaggart's main argument against the reality of time as nothing but "a philosophical 'howler'" – a logical blunder "of the same kind as the Ontological Argument for the Existence of God".³⁸

Broad's criticism of McTaggart has been very influential. It is exploited in Alexander Gunn's 1930 monograph, with its references to "the reality of changing objects", and to that "fundamental becoming" of the universe which "brings new events into being";³⁹ and its impact is still, or again, fully there in John Perry's paper "How Real Are Future Events?", given at the 2005 *Time and History* Kirchberg symposium.⁴⁰ Also, I would like to single out specifically the influence Broad had on Sellars. Recalling his time in Oxford in the mid-thirties, Sellars comes to compare G. E. Moore with Broad. "I had long felt", he tells us, "that, although C. D. Broad might not be clearer than Moore, nevertheless he had a more adequate grasp of the problems they shared. I now think", Sellars says, "that this can be traced to Broad's awareness of, and technical competence in, the scientific background of these problems."⁴¹ My impression is that, to some measure at least, it was under Broad's influence that Sellars developed his substantialist ontology of change, opposing the view that "when $S$ changes from being $\varphi$ to being $\psi$, $S$ must *really* consist of an event which is $\varphi$ and an event which is $\psi$ to be the terms for the relation *earlier than*". As Sellars saw the matter, "[t]hings couldn't *consist* of events, because events were the changes of things".⁴²

Let me conclude this section by briefly referring to an overlapping, but somewhat different, variety of anti-McTaggart argumentation – the ordinary-language variety – rather well represented by David Pears' 1956 essay "Time, Truth, and Inference". As Pears sees the matter, the paradoxes to which McTaggart's way of thinking about time leads are "the revenge which time takes on philosophers who deprive it of its proper means of expression, temporal verbs".⁴³ Focussing on the death of Queen Anne example, Pears discusses the EVENTS NEVER CHANGE argument,

---

37   Broad's "Ostensible Temporality" chapter I am here quoting from Richard M. Gale (ed.), *The Philosophy of Time*, p. 131.
38   *Ibid.*, p. 142.
39   Gunn, *The Problem of Time*, pp. 346 f.
40   John Perry, "How Real Are Future Events?", in F. Stadler and M. Stöltzner (eds.), *Time and History/Zeit und Geschichte*, Frankfurt/M.: ontos verlag, 2006, pp. 13–30.
41   Wilfrid Sellars, "Autobiographical Reflections", p. 284.
42   *Ibid.*, pp. 281 f.
43   David F. Pears, "Time, Truth, and Inference", in Antony Flew (ed.), *Essays in Conceptual Analysis*, London: Macmillan, 1956, p. 228.

finding that what McTaggart actually does is to turn, as it were, "the timeless shadows of the future (and the past) into contemporary things". McTaggart achieves this by making the timeless present tense, as Pears puts it, "refer to any time when really it refers to no time".[44] McTaggart's move relies on the misconception of *the eternity of truth*, a bizarre misconception which, Pears believes, might perhaps be psychologically explained by "a strong desire to know the future",[45] but is, nonetheless, logically untenable. There are no eternal truths, and there are no non-temporal facts. McTaggart was unable, or unwilling, to realize that "temporal predicates are unlike nontemporal predicates and that events are unlike things";[46] he was unwilling to yield to "the natural tendency of ordinary people to use temporal verbs". Had he done so, writes Pears, "his conclusion would have been not the unreality of time, but the unreality of timelessness".[47]

## Spurious Respectability

As Broad wrote, and indeed as Wittgenstein again and again lamented, philosophers, when doing philosophy, tend to be attracted to phoney language. Even so, the magic of McTaggart's systematically skewed syntax can by itself hardly explain the continuing influence his position exerts. As I suggested by way of introduction, the explanation is, rather, that this position has become systematically conflated with the Einstein–Minkowski conception of space-time, winning, thereby, undeserved esteem. There are innumerable places where McTaggart on the one hand, and relativity theory on the other, are mentioned in one breath; let me single out just a few. In the Einstein volume in the series *The Library of Living Philosophers*, published in 1949, the chapter by Kurt Gödel begins with a note referring to McTaggart's *Mind* paper. Peter Geach in his 1965 essay "Some Problems about Time" feels it his task to indicate that there is no *real* parallel between, on the one hand, the metaphysical genius McTaggart's conviction that time is an illusion, and on the other, the "view of time that is now widely held in one form or another. In its crudest form, this view makes time out to be simply one of the dimensions in which bodies are extended; bodies have not three dimensions but four. ... Since Einstein", Geach adds, "this sort of view has been very popular with philosophers

---

44  *Ibid.*, p. 232.
45  *Ibid.*
46  *Ibid.*, p. 230.
47  *Ibid.*, p. 235.

who try to understand physics and physicists who try to do philosophy."[48] Again, Hugh Mellor in his 1998 book *Real Time II* finds it necessary to argue against, as he puts it, the often-voiced falsehood that McTaggart's so-called "*B*-theory explains, and may even be entailed by, a key implication of Einstein's special theory of relativity, namely that the four dimensions of spacetime are in reality all alike".[49] Physicist Julian Barbour in his book *The End of Time*, published in 2000, aimed at demonstrating that time is but an illusion, notes that some ideas in McTaggart match his own thinking, although of course the latter's arguments "are purely logical and make no appeal to physics".[50] Very telling is the way Sider begins his 2001 book, bearing the subtitle *An Ontology of Persistence and Time*, by announcing that it "articulates and defends four-dimensionalism: an ontology of the material world according to which objects have temporal as well as spatial parts. ... The philosophy of time defended is the B-theory, the so-called 'tenseless theory of time'. ... The advent of Minkowski spacetime", writes Sider, "seems to have inspired much interest in [four-dimensionalism], although some versions of the doctrine predate Minkowski spacetime."[51] And to name a very recent publication: Sattig in his book *The Language and Reality of Time* opens by introducing in immediate succession first the McTaggartian notions of "A series" and "B series", and secondly the Minkowski–Einstein idea of spacetime.[52]

It is an historical coincidence that McTaggart's paper on "The Unreality of Time", published in the October 1908 issue of *Mind*, followed so closely upon Minkowski's famous *Raum und Zeit* talk, given at Cologne on September 21, 1908. But it is no more than a coincidence, having neither symbolic, nor indeed factual import. In his book *The Life and Philosophy of McTaggart*, Gerald Rochelle suggests that Einstein was aware of McTaggart's work.[53] This might easily be true, since Einstein probably had a look at Gödel's chapter in the volume I mentioned above. But Rochelle also suggests that McTaggart kept himself "in

---

48   P. T. Geach, "Some Problems about Time", in P. F. Strawson (ed.), *Studies in the Philosophy of Thought and Action*, London: Oxford University Press, 1968, pp. 175 f.
49   D. H. Mellor, *Real Time II* (1998), London: Routledge, 2006, p. 47.
50   Julian Barbour, *The End of Time: The Next Revolution in Our Understanding of the Universe* (1999), London: Phoenix, 2000, p. 343.
51   Theodore Sider, *Four-Dimensionalism: An Ontology of Persistence and Time*, Oxford: Clarendon Press, 2001, pp. xiii and 3.
52   Thomas Sattig, *The Language and Reality of Time*, Oxford: Clarendon Press, 2006, pp. 19–22.
53   Rochelle, *The Life and Philosophy of J. McT. E. McTaggart, 1866–1925*, Lewiston, NY: Edwin Mellen, 1991, p. xi.

touch with major scientific thinking", and "was most interested in Einstein's work on relativity"[54]. Rochelle offers no evidence for this, and I find it hard to believe. Rather, it is Broad who convinces me. This is what he writes in the 1933 "Preface" to his *Examination of McTaggart's Philosophy*: "I am inclined to think that McTaggart's complete lack of acquaintance with contemporary natural science was in certain respects a great advantage to him as a philosopher. The recent advances in physical theory have been so important and spectacular that they have only too obviously 'gone to the heads' of some eminent physicists, and have encouraged them and the public to believe that their pronouncements on technical philosophical problems, for which they have no special training or aptitude, are deserving of serious attention."

So the alleged McTaggart–Einstein connection is spurious. McTaggart's own logic is spurious. I think it is time for us to realize that McTaggart has, indeed, become a thing of the past. When did he become that? If I had the courage of my convictions, I would say that this happened as early as 1908, when he formulated, in the first passages of his *Mind* paper, THE A AND B SERIES STIPULATION. But certainly it happened by 1923 at the latest, when Broad's *Scientific Thought* saw the light of day. Or if you think that is still too harsh, then let us say it happened in 2005, when several papers at the Kirchberg *Time and History* symposium, most notably the neo-Broadian one given by John Perry, offered some decisive criticisms of McTaggart's position. And if you think I am too partisan, then let us look again, but this time from a different angle, at our much-discussed parallel, between McTaggart on the one hand, and Einstein–Minkowski on the other.

Wilfrid Sellars, in his 1962 paper "Time and the World Order", made the following remark: "The non-perspectival structure which, as realists, we conceive to underlie and support perspectival temporal discourse is, as yet, a partially covered promissory note the cash for which is to be provided not by metaphysics (McTaggart's C-series), but by the advance of science (physical theory of time)".[55] May I here make three comments. First, I do not think physics by itself can give us a theory of time; metaphysics, or more broadly, philosophy, will always play a role in synthesizing the concepts with which science grasps reality. Secondly, major discoveries in science evidently influence the way philosophers think: should the notion of time become really superfluous in science, the philosophy of time would

---

54 *Ibid.*, p. 186.
55 Wilfrid Sellars, "Time and the World Order", in *Minnesota Studies in the Philosophy of Science*, vol. III: *Scientfic Explanation, Space, and Time*, ed. by Herbert Feigl and Grover Maxwell, Minneapolis: University of Minnesota Press, 1962, p. 593.

clearly not remain unaffected. Thirdly, the "partially covered promissory note" Sellars refers to, today looks increasingly unlikely to be cashed; the scientific proof of a non-temporal universe does not seem to be forthcoming. What physics today tells us, forgive me the pun, is indeed a dark matter. Time may yet have a future.

## A Future for Time?

Leaving physics aside, but not losing sight of the metaphysical issue, let me now, by way of conclusion, enter the field of psychology, or, rather, of cognitive science.[56] Doubt as to the reality of time can arise because, in contrast to our sense of vision, hearing, touch, and so on, we do not seem to have a *sense of time*. A magisterial presentation of the issue was provided by William James in his *The Principles of Psychology*. "Let one sit with closed eyes", he wrote, "and, abstracting entirely from the outer world, attend exclusively to the passage of time". What do we perceive? Not, as it were, a "pure series of durations", but "[o]ur heartbeats, our breathing, the pulses of our attention, fragments of words and sentences that pass through our imagination".[57] Now heartbeats, breathing, attention, etc. all involve, as James was once more made aware by Hugo Münsterberg in 1889, the play of muscular tension and relaxation. According to Münsterberg, it is feelings in the muscles of the eye, the ear, and also muscles in the head, neck, etc., by which we estimate lengths of time. These perceptions of tension, "triggered off by real muscular contractions or by memories of the same", amount to *a direct sense of time*[58] – a physical encounter with time, we might say. As James puts it, "muscular feelings can give us the object 'time' as well as its measure".[59]

---

56 For a more detailed presentation of the argument of the present section see my paper "Film, Metaphor, and the Reality of Time", *New Review of Film and Television Studies*, vol. 7, no. 2 (June 2009), pp. 109–118.
57 James, *The Principles of Psychology*, vol. I, pp. 619 f.
58 Hugo Münsterberg, *Beiträge zur experimentellen Psychologie*, Heft 2: *Zeitsinn – Schwankungen der Aufmerksamkeit – Augenmass – Raumsinn des Ohres*, 1989, p. 20.
59 James, *op. cit.*, p. 637. I find it fascinating to compare these views by James with a passage he formulates in his *The Varieties of Religious Experience*: "There is a state of mind, known to religious men, but to no others", he there writes, "in which the will to assert ourselves and hold our own" has been displaced by a complete surrender to, and trust in, God. "In this state of mind, what we most dreaded has become the habitation of our safety... The time for tension in our soul is over, and that of happy relaxation, of calm deep breathing, of an eternal present, with no discordant future to be anxious about, has arrived" (William James, *The Varieties of Religious*

There exists a substantial research tradition which has demonstrated that to muscular sensations there correspond images of one's posture – schematic bodily images. And since the 1980s conceptual metaphor theory invites ever more detailed descriptions of how kinesthetic experiences give rise to so-called *image schemas*. An image schema, as Mark Johnson defines it, is "a recurring, dynamic pattern of our perceptual interactions and motor programs".[60] Now it is image schemata that give rise to a great number of fundamental metaphors. Recall that according to conceptual metaphor theory, metaphor is only incidentally "a device of poetic imagination and the rhetorical flourish", its essence consists in *"understanding and experiencing one kind of thing in terms of another"*[61]. Time is a much-discussed topic in conceptual metaphor theory. The essential finding is that "[m]ost of our understanding of time is a metaphorical version of our understanding of motion in space".[62] Earlier in this chapter I have referred to the "time-moving" and "ego-moving" metaphors. As Lakoff and Johnson interpret the matter, these metaphors are *"figure-ground reversals* of one another".[63] *Figure-ground reversal*: this brings us to gestalt psychology – and to film theory. In the 1930s, German-born psychologist Karl Duncker made the following discovery with respect to "figure" and "ground" in moving visual gestalts: the "figure" tends to move, the "ground" to stand still. When observers, say, stand on a bridge and look at the moving water, their perceptions will be veridical; but when they fixate the bridge, they and the bridge may be seen as moving along the river. Duncker explained the phenomenon by pointing out that "the object fixated assumes the character of the 'figure', whereas the nonfixated part of the field tends to become ground".[64] Psychologist of art and film theorist Rudolf Arnheim exploits this explanation to come to terms with a trivially well-known phenomenon in film. "[T]he setting photographed by the traveling camera", Arnheim points out, "is seen as moving across the screen,

---

*Experience: A Study in Human Nature* (The Gifford Lectures on Natural Religion Delivered at Edinburgh in 1901–1902), London: Longmans, Green & Co., 1902, p. 47).

60  Mark Johnson, *The Body in the Mind: The Bodily Basis of Meaning, Imagination, and Reason*, Chicago: The University of Chicago Press, 1987, p. xiv.

61  George Lakoff and Mark Johnson, *Metaphors We Live By*, Chicago: University of Chicago Press, 1980, pp. 3 and 5.

62  George Lakoff and Mark Johnson, *Philosophy in the Flesh: The Embodied Mind and Its Challenge to Western Thought*, New York: Basic Books, 1999, p. 139.

63  *Ibid.*, p. 149.

64  I am here quoting Duncker from Rudolf Arnheim, *Art and Visual Perception: A Psychology of the Creative Eye* (1954), exp. and rev. ed., Berkeley: University of California Press, 1974, p. 380.

mostly because the viewer receives the kinesthetic information that his body is at rest. Only in extreme cases, e.g., when enough of the entire environment is seen as moving, will the visual input overrule the kinesthetic." Normally however, when our "muscular experiences" tell us that we are at rest, it is "the street [that] is seen as moving. It appears to be actively encountering the spectator as well as the characters in the film, and assumes the role of an actor among actors."[65]

There is a very clear analogy here between, on the one hand, the time-moving metaphor and film's moving road, and, on the other, the ego-moving metaphor and the spectator's perception of moving along in the film's environment. Thinking of time as passing, and seeing the road pass by on the screen, appear to have the same motor background. And the perception of time passing is no more of an illusion than the perception of the road moving towards us, or receding behind us, on film. Our everyday metaphors of the flow of time are grounded in kinesthetic image schemata depicting reality. Contrary to what McTaggart believed, the common-sense view of the reality of time can be vindicated.

---

65   *Ibid.*, pp. 379 and 381.

# 3. Gombrich on Image and Time

There is a very close, indeed intrinsic, connection between the notions of image and time. Images are incomplete unless they are moving ones – unless, that is, they happen in time. On the other hand, time cannot be conceptualized except by metaphors, and so ultimately by images, of movement in space. That only the moving image is a full-fledged one is a fact that was fully recognized and articulated by Ernst Gombrich.[1] And of course Gombrich entertained, and argued for, a rich and well-balanced view of the relationships between pictorial and verbal representation. An antidote to the unholy influence of Goodman,[2] Gombrich deserves to be rediscovered, or indeed discovered, in particular in Germany, as the figure whose work, complemented by that of Rudolf Arnheim[3] and possibly by

---

1   I had been unaware of this particular aspect of Gombrich's work when I wrote my paper "The Picture Theory of Reason" (given at the 2000 International Ludwig Wittgenstein Symposium, Kirchberg am Wechsel, published in Berit Brogaard and Barry Smith, eds., *Rationality and Irrationality*, Wien: öbv-hpt, 2001), a paper in which I noted that mental imagery appears to be a matter of dynamic, rather than static, pictorial representations, that still images are, psychologically speaking, but *limiting cases* of dynamic ones, and that, with the development of twentieth-century visual culture, the same seems to have become the case with regard to pictures in the world around us, too – think of film and video. On the other hand, in that paper I referred to the Oxford philosopher H. H. Price, who in his 1953 book *Thinking and Experience* (cf. chapter 1, note 58 above) had put forward the idea that while static images stand in need of interpretation because of their systematic ambiguity, "cinematographic" images go a long way towards being unambiguous.
2   Although Nelson Goodman's *Languages of Art* was very much inspired by Gombrich, the latter, as I noted in "The Picture Theory of Reason", had in the years following upon the publication of his *Art and Illusion* moved closer to a naturalistic account of images, coming to see in Goodman but an extreme relativist or conventionalist.
3   Gombrich and Arnheim were rivals, and the former's dubious praise of the latter in his *Art and Illusion: A Study in the Psychology of Pictorial Representation*, London: Phaidon Press, 1960, p. 22, was reciprocated with some biting criticisms by Arnheim in several reviews he wrote of Gombrich (on *Art and Illusion*, in *Art Bulletin* 44, March 1962; on *The Sense of Order*, in *The New Republic*, 10 March 1979; and on the collection *The Image and the Eye*, in *Times Literary Supplement*, 29 October 1982).

that of Hans Belting,[4] is ideally suited to providing a founding paradigm for a truly successful philosophy of images.

## Discovering Gombrich

To this day, Gombrich is primarily known as the author of the book *Art and Illusion*, first published in 1960. Now although in that book, as I will attempt to show in this chapter, the beginnings of what we can call Gombrich's philosophy of images are certainly present, it was a number of studies written in the 1960s

---

However, seen from today's perspective, the parallels in the work of the two seem to be much more important than the differences (this is the view taken also by Ian Verstegen, in his "Arnheim and Gombrich in Social Scientific Perspective", *Journal for the Theory of Social Behaviour*, vol. 34, no. 1, 2004). Two ideas which are significantly more marked in the work of Arnheim than in that of Gombrich are the primordiality of the pictorial, and the possibility of generic images; there can be no doubt that here Gombrich will gain by being supplemented by Arnheim. On the other hand, a seemingly promising avenue that might appear to lead to a better understanding of the similarities between Gombrich and Arnheim, namely the issue of their both being indebted to the work of Wolfgang Köhler, turns out to be a blind alley. Arnheim studied with Köhler and with Max Wertheimer, Köhler having earlier served as subject for Wertheimer's experiments on apparent movement, and it is obvious that Arnheim's notions about vision in general and the moving image in particular are very much rooted in the Wertheimer–Köhler Gestalt tradition. But while Gombrich actually took up a university course delivered by Köhler in Berlin in the 1930s, met the latter in Princeton after the war, and referred repeatedly to him in his writings beginning with *Art and Illusion*, the two were (some contrary allusions notwithstanding) never close, and Köhler's ideas left no real trace in Gombrich's work.

4  Gombrich's occasional references to the *mask* (see e.g. his "Visual Discovery through Art", *Arts Magazine*, November 1965, repr. in James Hogg, ed., *Psychology and the Visual Arts*, Harmondsworth, Middlesex: Penguin, 1969, p. 227, and esp. his "The Mask and the Face: The Perception of Physiognomic Likeness in Life and Art", in Gombrich et al., *Art, Perception, and Reality*, Baltimore: The Johns Hopkins University Press, 1972) as well as to the "art of makeup" as "one of the oldest forms of visual art" (see his "The Evidence of Images", in C. S. Singleton, ed., *Interpretation, Theory and Practice*, Baltimore: Johns Hopkins University Press, 1969) do certainly not add up to an *anthropology of images* in the sense of Belting. See the latter's *Bild-Anthropologie: Entwürfe für eine Bildwissenschaft*, München: Wilhelm Fink Verlag, 2001.

and 1970s in which that philosophy was actually elaborated.[5] Let me here list the ones I consider most important. 1964 saw the appearance of the essay "Moment and Movement in Art",[6] of central importance to the topic of image and time. The 1965 paper "Visual Discovery through Art",[7] presented by Gombrich as a taking stock once more of, and a formulating of some afterthoughts on, the issues dealt with in *Art and Illusion*, is a major step forward in dealing with the problems of pictorial realism, generic images, and visual context. In the lengthy study "The Evidence of Images", published in 1969, where the tone is set by a quote from Ulric Neisser referring to Brentano, Bergson, and James, with Neisser stressing that "the mechanisms of visual imagination are continuous with those of visual perception",[8] Gombrich adds substantial new material to his discussion in *Art and Illusion* of visual perception as being dependent on movement. The paper "The Mask and the Face",[9] Gombrich's 1970 Thalheimer Lecture, recapitulates ideas from the chapter on caricature in *Art and Illusion*, but also represents another significant move towards coming to terms with the topic of time and image. The essay "The Visual Image", written for a *Scientific American* 1972 special issue on communication, argues for the joint exploitation of the media of word and image, but arrives at the momentous formulation that the "real value of the image ... is its capacity to convey information that cannot be coded in any other way".[10] 1972 saw Gombrich's first direct attack on Goodman,[11] the former's main contentions here being that "Goodman appears to think that the eye must be strictly stationary" whereas "no stationary view can give us complete information", and also that the pictorial technique of *perspectival representation* reflects something essentially

---

5  In taking this view of the matter, I feel encouraged by a conversation I had in 2009 with Richard Woodfield, creator of the online Gombrich Archive, Honorary Senior Research Fellow in the Department of Art History at the University of Glasgow. I am deeply indebted to Woodfield for his continuous and unfailing help in extending my knowledge of Gombrich.
6  E. H. Gombrich, "Moment and Movement in Art", *Journal of the Warburg and Courtauld Institutes*, XXVII (1964), pp. 293–306.
7  Cf. note 4 above.
8  "The Evidence of Images" (cf. note 4 above), p. 40.
9  Cf. note 4 above.
10  E. H. Gombrich, "The Visual Image", *Scientific American*, vol. 227, no. 3, September 1972, p. 87.
11  E. H. Gombrich, "The 'What' and the 'How': Perspective Representation and the Phenomenal World", in Richard Rudner and Israel Scheffler (eds.), *Logic & Art: Essays in Honor of Nelson Goodman*, Indianapolis: Bobbs-Merrill, 1972.

natural and objective – it does not need to be learned to be decoded.[12] The second, devastating, attack came six years later, with Gombrich's paper "Image and Code: Scope and Limits of Conventionalism in Pictorial Representation",[13] vindicating the common-sense idea of pictures as natural signs, and explicating the controversial concept of *resemblance* by that of *equivalence of response*.[14] As Gombrich here momentously puts it: "the images of Nature, at any rate, are not conventional signs, like the words of human language, but show a real visual resemblance, not only to our eyes or our culture but also birds or beasts".[15] Finally, the paper "The Arrested Image and the Moving Eye", published in 1980, further pursued the crucial issue of vision and mobility, stressing that the "perception of movement is different in character from the inspection of a static scene".[16]

My impression is that the ideas put forward in these writings have never been fully absorbed by Gombrich's readers. Let me here give a few examples, perhaps somewhat random, but together, I believe, adding up to a picture. The prominent American film theorist David Bordwell is definitely an admirer of Gombrich. In his 1997 book *On the History of Film Style*, he speaks of Gombrich's "scintillating career"[17] and sees himself as "asking the cinematic counterpart of the question that opens E. H. Gombrich's *Art and Illusion*: Why does art have a history?".[18] His earlier book *Narration in the Fiction Film*, too, is very much written in the wake of Gombrich; Bordwell here not only makes numerous references to *Art and Illusion*, stressing, mainly, the element of convention and construction in

---

12  *Ibid.*, pp. 133, 136 and 148.
13  Delivered at a symposium in 1978, published in Wendy Steiner (ed.), *Image and Code*, Ann Arbor: University of Michigan Press, 1981.
14  *Ibid.*, pp. 11 and 17.
15  *Ibid.*, p. 21. This is the stance Arnheim refers to in his *Times Literary Supplement* review (cf. note 3 above) when he writes that here "Gombrich rises to the defence of the visual image and its inherent truthfulness, to which even animals respond – an image shaped by simplification and abstraction, to be sure, and by the conventions of pictorial styles, but nature's message nevertheless. ... It is from this secure basis that Gombrich's future work should be able to proceed."
16  E. H. Gombrich, "Standards of Truth: The Arrested Image and the Moving Eye", in W. J. T. Mitchell (ed.), *The Language of Images*, Chicago: University of Chicago Press, 1980, p. 206.
17  David Bordwell, *On the History of Film Style*, Cambridge, MA: Harvard University Press, 1997, p. 150.
18  *Ibid.*, p. 3.

comprehending images,[19] but draws also on several other studies by Gombrich, in particular on the paper "Image and Code", saying: "There is, Gombrich points out, a continuum between natural skills and acquired ones. It seems evident that the ability to comprehend 'scientific' perspectival images is much more easily acquired than, say, the ability to read a language. Perhaps perspectival cues build upon some natural skills, such as the organism's ability to detect surfaces and edges."[20] However, his familiarity with "Image and Code" notwithstanding, Bordwell still attributes to Gombrich the position that "all images are inherently ambiguous"[21] – even though, to recall, it is in "Image and Code" that Gombrich makes the strongest case for the position that images can function as unequivocal natural signs.

A recent book by the renowned philosopher of science Bas van Fraassen, *Scientific Representation: Paradoxes of Perspective*, cannot but address some questions that had been at the centre of interest in Gombrich's work. Van Fraassen mentions Gombrich only once, though at the very beginning of the book,[22] but in an incidental context. He takes from *Art and Illusion* a passage Gombrich quotes on Phidias and Alcamenes competing with each other,[23] with Phidias recognizing what Alcamenes did not, that in art distortion might be necessary to achieve faithful rendering. Van Fraassen then goes on to discuss caricature and misrepresentation – a favourite topic of Gombrich's – stressing that "likeness" or "resemblance" are elusive notions; that resemblance is always *selective*.[24] But this is a blunder, one that van Fraassen might have avoided by paying closer attention to Gombrich. As the latter had shown in detail in "Image and Code", the notion of resemblance can be derived from that of visual equivalence. It is not resemblance that is selective, but *equivalence*. Resemblance *is* selective equivalence.[25]

We should cast a glance on the German scene, too. Let me first single out Gottfried Boehm and Oliver Scholz. Introducing his 1985 talk "Image and Time",

---

19   David Bordwell, *Narration in the Fiction Film*, Madison, WI: The University of Wisconsin Press, 1985, p. 33 (cf. note 16 on p. 343) and p. 102 (cf. note 9 on p. 347).
20   *Ibid.*, p. 107, note 24 here referring to pp. 17–21 of "Image and Code", in Wendy Steiner (ed.), *Image and Code*.
21   *Ibid.*, p. 102.
22   Bas C. van Fraassen, *Scientific Representation: Paradoxes of Perspective*, Oxford: Clarendon Press, 2008, pp. 12 f.
23   *Art and Illusion*, p. 162.
24   *Scientific Representation*, pp. 18, 33, 57, and *passim*.
25   "Image and Code", pp. 17 and 21.

Boehm points to his long-standing interest in the problem of time.[26] In the talk, he very briefly mentions Gombrich's "Moment and Movement in Art",[27] and later makes a passing reference to *Art and Illusion* in a note.[28] When one thinks of the breadth and depth of Gombrich's work on the problems of image, movement, and time, Boehm's parsimoniousness in exploiting the former's results seems somewhat surprising. And quite odd is the way Scholz treats Gombrich in his *Bild, Darstellung, Zeichen*. He designates *Art and Illusion* as an epoch-making investigation,[29] and lists Gombrich's work (together with the writings of Barthes and Goodman) as one of the "initial ignitors" of the interest in pictorial representation,[30] but then mentions him only very occasionally, mostly in slighting terms, and with practically no reference to his post-1960 studies.[31]

In the contemporary German reception of Gombrich, a definitely exceptional role is played by Klaus Sachs-Hombach, who in his book *Das Bild als kommunikatives Medium* provides an illuminating and balanced picture of the former's results. Gombrich's real contribution to a theory of images, stresses Sachs-Hombach, consists in his showing that resemblance and cultural conditioning both play a role in pictorial perception.[32] Gombrich is not a conventionalist in the sense of Goodman, but nor does he believe that aiming at resemblance necessarily involves the attempt to set up an illusion.[33] According to Gombrich, it is significant that images created by nature will fulfil their function without displaying perfect likeness. As Sachs-Hombach puts it: "the success of imitations – and of the various forms of mimicry in the animal and plant world – does not at all depend on the images being as naturalistic as possible; on the contrary, it is schematized representations that are, as a rule, the most suitable, with a rough rendering of size and form, display-

---

26  Gottfried Boehm, "Bild und Zeit", in Hannelore Paflik (ed.), *Das Phänomen Zeit in Kunst und Wissenschaft*, Weinheim: VCH, 1987, p. 1, starred note.
27  *Ibid.*, p. 5.
28  *Ibid.*, p. 8, note 13.
29  Oliver R. Scholz, *Bild, Darstellung, Zeichen: Philosophische Theorien bildlicher Darstellung*. 2nd, completely rev. ed., Frankfurt/M.: Klostermann, 2004, p. 2.
30  *Ibid.*, p. 4.
31  The single exception is a reference, in note 51 on p. 168, to Gombrich's 1961 essay "How to Read a Painting".
32  Klaus Sachs-Hombach, *Das Bild als kommunikatives Medium: Elemente einer allgemeiner Bildwissenschaft*, Köln: Herbert von Halem Verlag, 2003, pp. 135–139.
33  *Ibid.*, p. 194

ing some essential species-specific characteristic".[34] Man-made images, too, might well carry definite meanings by themselves – without the help of conventions; this is especially true when it comes to moving images. "With the temporal dimension of film", writes Sachs-Hombach, "there occurs a disambiguation of what is represented – dispelling many uncertainties, and leading to a more immediate, perception-like, recognition of pictorial content".[35] In Gombrich's work, Sachs-Hombach clearly suggests, meaning, image, and time are closely bound up with each other.

# Word and Image

While recognizing the communicative potential of images, Gombrich is fully aware, as I indicated earlier, of the role of *language* in pictorial representation – of the complex interrelationships between word and image. In *Art and Illusion*, he was fond of talking of the "linguistics of the visual image", or the "language of art",[36] but this was but a metaphoric way of expressing himself: what he had in mind were the vocabulary and grammar, if you like, of *pictorial schemata*, acquired graphic formulas.[37] The real issue of image and word is the one Gombrich introduces at the beginning of *Art and Illusion* with the reference that it was his early, joint research with Ernst Kris "into the problem of caricature" which first confronted him with "the question of what is involved in accepting an image as a likeness".[38] The problem of likeness in caricature is of course just a special case of the problem of likeness in images: in portraits, but also, say, in landscapes. Trivi-

---

34   *Ibid.*, p. 268, referring to Gombrich's "Visual Discovery through Art", in Hogg (ed.), *Psychology and the Visual Arts* (cf. note 4 above), pp. 226 f.
35   *Ibid.*, p. 229.
36   *Art and Illusion*, p. 7.
37   "Everything points to the conclusion", writes Gombrich, "that the phrase 'the language of art' is more than a loose metaphor, that even to describe the visible world in images we need a developed system of schemata" (*Art and Illusion*, p. 76). What Gombrich here means is clear – he applies a metaphor, even if not a "loose" one – but still it is instructive to look at another passage in *Art and Illusion*, where he makes a reference to Hogarth, in whose view the artist "should 'learn the language' of objects and 'if possible find a grammar to them'." To which Gombrich adds: "In other words, [the artist] should stock his mind well with what we called 'schemata'" (*ibid.*, p. 295). This is the sense in which, in the concluding passage of the chapter on caricature, Gombrich says: "Wherever the artist turns his gaze he can only make and match, and out of a developed language select the nearest equivalence" (*ibid.*, p. 303).
38   *Ibid.*, p. ix.

ally, two-dimensional pictures, whether line drawings, paintings, or photographs, colour or black-and-white, are not at all like what they represent. However, as Gombrich points out, there are ways to create, and to discern, certain *identities*, or equivalences, that do indeed pertain to the image and its object. "The invention of portrait caricature", he writes, "presupposes the theoretical discovery of the difference between likeness and equivalence."[39] Or, more generally: "All artistic discoveries are discoveries not of likenesses but of equivalences which enable us to see reality in terms of an image and an image in terms of reality. And this equivalence never rests on the likeness of elements so much as on the identity of responses to certain relationships."[40] It is here we find the germ of the idea that will surface in its fully developed form in the paper "Image and Code", in 1987. Equivalences meet the eye, but the pictorial information they convey might not be interpretable in the absence of verbal pointers such as labels and captions.[41] Only with its label added will Constable's painting of Wivenhoe Park "tell us a good many facts about that country-seat in 1816";[42] only together with the caption "What have you done with Dr. Millmoss?" will the drawing by James Thurber ("with much charm and humour") recount its sad message (Figure 1).[43] And only the combination of drawing and text creates the specific experience provided by the Töpffer variety of the "picture story", a precursor of the comic strip.[44] However, the often crucial role of verbal explanations notwithstanding, images also have to speak for themselves. In the 1962 preface to the second edition of *Art and Illusion*, Gombrich stresses that "the undeniable subjectivity of vision does not preclude objective standards of representational adequacy", and points to "the dissatisfaction which certain periods of

---

39  *Ibid.*, p. 290.
40  *Ibid.*, p. 292.
41  Cf. esp. *ibid.*, pp. 59 f., 64 and 77.
42  *Ibid.*, p. 252.
43  *Ibid.*, 302.
44  *Ibid.*, pp. 284 f. The passages Gombrich here quotes from Töpffer are instructive: "There are two ways of writing stories, one in chapters, lines, and words, and that we call 'literature', or alternatively by a succession of illustrations, and that we call the 'picture story'. ... The picture story ... has always exercised a great appeal. More, indeed, than literature itself, for besides the fact that there are more people who look than who can read, it appeals particularly to children and to the masses... With its dual advantages of greater conciseness and greater relative clarity, the picture story, all things being equal, should squeeze out the other because it would address itself with greater liveliness to a greater number of minds."

Western civilization felt with images that failed to look convincing".[45] Here, the invention of "the art of perspective" aiming at a "correct equation" was a major step forward.[46]

*Figure 1:*   James Thurber, *"What have you done with Mr. Millmoss?"* Compare: Gombrich, *Art and Illusion*, p. 302

Gombrich returns to this last topic in the paper "Visual Discovery through Art". It is not at all the case, he writes, that mathematical perspective represents "only ... a 'convention', a fortuitous code that differs from the way we really see the world". As he puts it: "we know very well when a picture looks 'right'. A picture painted according to the laws of perspective will generally evoke instant and effortless recognition. It will do so to such an extent that it will in fact restore the feeling of reality."[47] The felt need leading to the invention of perspective in the 15th century was of a religious nature: the demand for "the plausible narration of sacred events. ... The closer the code came to the evocation of a familiar reality the more easily could the faithful contemplate the re-enactment of the story and identify the participants."[48]

The issue of word and image very much takes centre stage in Gombrich's essay written for the 1972 *Scientific American* survey on communication. "Ours is a visual age", Gombrich here writes by way of introduction. "We are bombarded with pictures from morning till night. ... No wonder it has been asserted that we

---

45  E. H. Gombrich, *Art and Illusion: A Study in the Psychology of Pictorial Representation*, 2nd ed., London: Phaidon Press, 1962, p. xi.
46  *Art and Illusion*, 1960, p. 217, cf. also p. 279.
47  "Visual Discovery through Art", 1965 (cf. note 4 above), p. 222.
48  *Ibid.*, pp. 223 f.

are entering a historical epoch in which the image will take over from the written word. In view of this claim it is all the more important to clarify the potentialities of the image in communication, to ask what it can and what it cannot do better than spoken or written language."[49] Images are inferior to language when it comes to logical relations, tense, and modality. As Gombrich puts it, "the visual image ... unaided ... altogether lacks the possibility of matching the statement function of language".[50] To be understood fully, the image has to be embedded in cultural conventions and complemented by verbal guides. "The chance of a correct reading of the image", writes Gombrich, "is governed by three variables: the code, the caption and the context. ... Jointly the media of word and image increase the probability of a correct reconstruction."[51] Gombrich prints the mosaic of a dog found at the entrance of a house in Pompeii (Figure 2). The mosaic has the inscription *Cave Canem* ("Beware of the Dog"). Without the inscription, Gombrich points out, the message intended to be communicated by the mosaic would be unclear.

*Figure 2:*   *"Cave Canem", mosaic from Pompeii.*
  *Compare Gombrich, "The Visual Image", p. 85*

---

49   "The Visual Image" (cf. note 10 above), p. 82.
50   *Ibid.*
51   *Ibid.*, p. 86. To which Gombrich adds: "[the] mutual support of language and image facilitates memorizing. The use of two independent channels, as it were, guarantees the ease of reconstruction."

On the other hand, images can carry information no verbal description will provide, images as natural signs easily possess a kind of primordial power, "organisms are 'programmed' to respond to certain visual signals in a way that facilitates survival",[52] images *affect* us. This way of looking at the issue becomes especially pronounced in the paper "Image and Code". The Pompeii mosaic is here again reproduced, with Gombrich emphasizing that in order to understand that the dog depicted looks menacing, we do not have to learn specific stylistic conventions; and that, in particular, "we do not have to acquire knowledge about teeth and claws in the same way in which we learn a language".[53] Indeed, even animals respond to images. As Gombrich, arguing against Goodman, puts it: "Images have always been used to attract or frighten animals. What else is a decoy duck or the angler's bait than an image securing the reaction of another creature? … the fish which snaps at the artificial fly does not ask the logician in what respect it is like a fly and in what unlike."[54]

Towards the end of the paper "Visual Discovery through Art", Gombrich returns to the ambiguous duck/rabbit figure he had discussed in *Art and Illusion*. We can prompt alternate readings, he notes, depending on captions, i.e., on verbal descriptions, "but it might be even more effective to impose one of these readings through visual means". As he puts it, though he has not made experiments, he would predict that one could "bring about a transformation merely by changing the visual context", either *spatially*, by drawing a typical duck or rabbit habitat around the ambiguous figure, or *temporally*, by showing a subject "a series of pictures", of ducks or rabbits, "before projecting the ambiguous image".[55] The idea of a series of pictures, of images changing temporally, is paramountly important – and one which takes me to the remaining two sections of the present chapter, the sections on image and movement, and on movement and time.

---

52   *Ibid.*, p. 85.
53   "Image and Code", p. 20. Referring to John M. Kennedy's *A Psychology of Picture Perception: Images and Information* (San Francisco: Jossey-Bass, 1974), Gombrich, some pages earlier, makes the remark: "the widespread view has recently been challenged that the conventional elements in photographs bar naive subjects such as unsophisticated tribesmen from reading them. At any rate it appears that learning to read an ordinary photograph is very unlike learning to master an arbitrary code. A better comparison would be with learning the use of an instrument. It is quite possible that many tribesmen who are handed a photograph will not know at first what to do with it, or how they are expected to look at it, but I assume their reaction would be similar if they were handed a pair of binoculars. You have to learn to use it" ("Image and Code", p. 16).
54   *Ibid.*, p. 20.
55   "Visual Discovery through Art", p. 235.

# Image and Movement

As I indicated earlier, the intrinsic connections between vision and movement have been of central interest to Gombrich throughout his career. In *Art and Illusion* he pointed to "the total ambiguity of one-eyed static vision",[56] stressing the importance "the test of movement" has when it comes to dissolving uncertainties in our visual field. "Whenever we do not quite trust our eyes or want additional information", he wrote, "we shift our head slightly and watch the relative change of position."[57] The illusion created by a still life tends to disappear as soon as we move; in the elimination of false visual guesses, movement, our own and that of objects, plays a vital part.[58] Images can strike us as unnatural when the aspect of movement is missing. "What we experience as a good likeness in a caricature, or even in a portrait", wrote Gombrich,

> is not necessarily a replica of anything seen. If it were, every snapshot would have a greater chance of impressing us as a satisfactory representation of a person we know. In fact only a few snapshots will so satisfy us. We dismiss the majority as odd, uncharacteristic, strange, not because the camera distorts, but because it caught a constellation of features from the melody of expression which, when arrested and frozen, fails to strike us in the same way the sitter does. For expression in life and physiognomic impression rest on movement no less than on static symptoms.

By contrast, "the snapshot ... arrests movement and fixes it for ever".[59] The challenge for art, then, is to create, in static images, the suggestions of movement, to catch, as Velázquez did in the *Hilanderas* (Figure 3), "the so-called 'stroboscopic effect', the streaking after-image that trails its path across the field of vision when an object is whizzing past", an effect the suggestion of which today "belongs to the commonplace language of the cartoonist or comic-strip artist. There is hardly a picture narrative in which speed is not conveniently rendered by a few strokes which act like negative arrows showing where the object has been a moment before."[60]

---

56  *Art and Illusion*, p. 330.
57  *Ibid.*, p. 232.
58  *Ibid.*, pp. 234, 179 and 277.
59  *Ibid.*, pp. 292 f.
60  *Ibid.*, pp. 191 f. The phrase "negative arrows", on p. 192, refers back to what Gombrich wrote on the previous page: "It appears that if you show an observer the images of a pointing hand or arrow, he will tend to shift its location somehow in the direction of the movement. Without this tendency of ours to see potential movement in the form of anticipation, artists would never have been able to create the suggestion of speed in stationary images."

*Figure 3:* Velázquez, "Hilanderas".
Compare Gombrich, *Art and Illusion,* p. 192

Gombrich had a great deal of respect for the psychologist J. J. Gibson. In the preface to *Art and Illusion*, he acknowledged his indebtedness to the latter's 1950 book *The Perception of the Visual World*; in the paper "The Evidence of Images", he comes to terms with Gibson's *The Senses Considered as Perceptual Systems*, published in 1966. Gibson's work, Gombrich here suggests, "has initiated what may be called a Copernican revolution in the study of perception". But Gombrich is reluctant to accept Gibson's "radical separation between the interpretation of pictures and the perception of the world". Gibson might, for instance, be right in assuming that "in walking or driving along a road we would have information of a very different order from what the snapshot gives us, and that we thus could perceive the invariant shape of the road, the houses, and the texture of the road without ambiguity"; however, Gombrich objects, it is "not sure how far our capacity to process this information would ever go". But he of course wholeheartedly endorses Gibson's basic position, according to which "visual perception is geared to movement". Gombrich grants Gibson that "the static view of a room through a stationary eye allows of many interpretations", and accepts the latter's view that as soon as we change our position, the "transformation of the optic array" becomes unequivocal: "there is one and only one configuration which fits it. What matters in real life is not that textbook abstraction, the stationary image on one retina, but the succession of stimuli which we experience as we are walking toward a room."[61]

Reference is again made to *The Senses Considered as Perceptual Systems* in the paper "The Mask and the Face". Thanks to Gibson's work in the psychology of perception, Gombrich writes, "we have become increasingly aware of the decisive role which the continuous flow of information plays in all our commerce with the visible world".[62] The idea of *flow*, as opposed to that of static permanence, here proves to be a significant one; it makes Gombrich arrive at some momentous observations. The *snapshot*, he writes, has not only "transformed the portrait", it has also "made us see the problem of likeness more clearly than past centuries were able to formulate it. It has drawn attention to the paradox of capturing life in a still, of freezing the play of features in an arrested moment of which we may never be aware in the flux of events."[63] To which he adds a crucial passage:

> if the film camera rather than the chisel, the brush, or even the photographic plate had been the first recorder of human physiognomies, the problem which language in its wisdom calls "catching a likeness" would never have obtruded itself to the same extent on our awareness. The film shot can never fail as signally as the snapshot can, for even if it catches a person blinking or sneezing the sequence explains the resulting grimace which the corresponding snapshot may leave uninterpretable. Looked at in this way, the miracle is not that some snapshots catch an uncharacteristic aspect, but that both the camera and the brush can abstract from movement and still produce a convincing likeness not only of the mask but also of the face, the living expression.[64]

Gibson is once more invoked in the paper "The 'What' and the 'How'". It is Gibson's approach, Gombrich here suggests, that most fully explains how "our own movement", with our "phenomenal world" in constant motion, produces "a fluctuating succession of images", and how this "flux of events" is captured in a "stream of information". No wonder *movies* tend to be more immediately realistic than stills: "In the motion picture the rapid enlargement of an object can make us duck."[65] Again, it is very much in Gibson's spirit that Gombrich discusses, in his paper "Image and Code", the basic element of the two-dimensional image, the *outline*. "It has often been said", Gombrich here writes, "that the outline is a convention because the objects of our environment are not bounded by lines. ... yet ... [t]hings in our

---

61 "The Evidence of Images" (cf. 4 above), pp. 45, 47 and 44.
62 "The Mask and the Face" (cf. note 4 above), pp. 16 f.
63 *Ibid.*, p. 16
64 *Ibid.*, p. 17.
65 "The 'What' and the 'How'" (cf. note 11 above), pp. 137 and 139.

environment are indeed clearly separated from their background, at least they so detach themselves as soon as we move. The contour is the equivalent of this experience; it indicates what would happen if the image were not a still but would change, as the world around us usually does."[66]

In his essay "Standards of Truth: The Arrested Image and the Moving Eye", Gombrich recalls how Gibson came to derive his theories "from his wartime work when he investigated the visual information available to a pilot landing at high speed. It is not a static image which gives the pilot the required estimate of the distance and position of the runway but the flow of information he receives, the sequence of transformations all around which show him across these rapid changes the invariants of the lay of the land, invariants he must pick up if he is to survive."[67] To which Gombrich later in the essay adds: "peripheral vision is extremely sketchy in the perception of shapes and colours but very responsive to movement. We are aware of any displacement in the medley of forms outside the foveal area and ever ready to focus on such an unexpected intrusion. Once we have done so we can track the moving object without letting it go out of focus, while the rest of the field of vision recedes from our awareness. There is no means of conveying this experience in a stationary display."[68] The position Gombrich here takes has been first outlined by the 19th-century physicist, physiologist, and psychologist Hermann von Helmholtz. "Thanks to the mobility of the eye", Gombrich quotes Helmholtz, "it is possible to examine carefully every point of the visual field in succession. Since in any case we are only able to devote our attention at any time to one object only, the one point clearly seen suffices to occupy it fully whenever we wish to turn to details; on the other hand the large field of vision is suitable, despite its indistinctness, for us to grasp the whole environment with one rapid glance and immediately to notice any novel appearance on the margin of the field of vision."[69]

Analyzing the "artificial situation of arrested movement", in the paper "The Mask and the Face" Gombrich once more points out that when it comes to understanding images, it is, precisely, *movement* that primarily assists us "in confirming or refuting our provisional interpretations or anticipations". As a consequence, "our reading of the static images of art is particularly prone to large variations and contradictory interpretations". It is, he says, the "dimension of time, above all, we lack in the interpretation of a still". By contrast, in "real life" we are invariably aided

---

66  "Image and Code", p. 17.
67  "Standards of Truth" (cf. note 16 above), p. 188.
68  *Ibid.*, p. 206.
69  *Ibid.*, p. 204.

"by the effect of movement in time".[70] Pictorial meaning cannot be discussed without reference to movement; and the topic of the moving image necessarily leads to the topic of time.

## Movement and Time

Gombrich provides a focussed discussion of the topic of time in his essay "Moment and Movement in Art". There are of course recurring references to the issue in *Art and Illusion*,[71] as well as some hints in the study "The Evidence of Images",[72] but it is in this 1964 essay that he presents what in fact amounts to the outlines of a psychology of time, and indeed of a *philosophy* of time. The way in which "the problem of the passage of time in painting was traditionally posed", he writes, has "doomed the answers to relative sterility"[73] precisely because it was based on a mistaken view of the nature of time, the view presupposing the existence of a *punctum temporis*, a view formulated by James Harris in his influential *Three Treatises* (1744), foreshadowed by Shaftesbury in the *Characteristics* (1714) when speaking of the "determinate Date or Point of Time", of the "single Instant", the artist has to choose when depicting a certain event in a narrative, and taken over by Lessing in his *Laocoon*, writing: "Painting can ... only represent a single

---

70 "The Mask and the Face", pp. 31 f.
71 The most interesting ones perhaps on p. 292, in the chapter "The Experiment of Caricature", where Gombrich remarks that "art has to compensate for the loss of the time dimension by concentrating all required information into one arrested image", and, paraphrasing Houbraken, puts the question: "how are you to copy rapid movement, running, flying, jumping? These will be over before you ever put pen to paper. ... how are you to copy ... the 'expression of human passions'? ... genuine expression ..., too, happens in time." The first footnote in "Moment and Movement in Art" is connected to some "relevant observations scattered throughout the literature" on the "strangely neglected" problem of "time and the representation of movement", listing, among others, Arnheim's *Art and Visual Perception*, chapter VIII, but also Gombrich's own *Art and Illusion*, referring to the book's index *sub verbo* "movement".
72 In a somewhat Bergsonian tone, Gombrich here writes: "We process the successive frames of the film as information about movement... We see movement, not a succession of stills." Some paragraphs later he proposes to rely on "the hypothesis that the isolation and reconstruction of an object is an operation in time which for all its rapidity is certainly complex", adding: "the reading of a picture is indeed a reaction in time" ("The Evidence of Images", pp. 57, 61 and 63).
73 "Moment and Movement in Art" (cf. note 6 above), p. 293.

moment of action and must therefore select the most pregnant moment which best allows us to infer what has gone before and what follows."[74]

That such an instant, such a moment, does not seem to exist, Gombrich first points out by recounting the story of early photography. Muybridge's snapshots of galloping horses did not suggest the melody of movement painters had believed to see; the instantaneous photograph looked unreal. It is not by chance, then, that the "so-called 'stills' which we see displayed outside cinemas and in books on art of the film are not, as a rule, simply isolated frames from the moving picture enlarged and mounted. They are specially made and very often specially posed on the set, after a scene is taken." It is surely true, Gombrich writes, that "we never see what the instantaneous photograph reveals, for we gather up successions of movements, and never see static configurations as such. And as with reality, so with its representation. The reading of a picture again happens in time… … it takes … time to sort a painting out. We do it … by scanning it with our eyes. Photographs of eye movements suggest [how] the eye probes and gropes for meaning…"[75]

On a philosophical level, Gombrich suggests that we are actually begging the most important question "when we ask what 'really happens' at any point of time". For we

> therewith assume that what Harris called a *punctum temporis* really exists, or, more radically, that what we really perceive is the infinite sequence of such static points in time. Once this is conceded the rest follows, at least with the demand for mimesis. Static signs, the argument runs, can only represent static moments, never movements which happen in time. Philosophers are familiar with this problem under the name of Zeno's paradox… Logically the idea that there is a "moment" which has no movement and can be seized and fixed in this static form by the artist, or, for that matter, by the camera, certainly leads to Zeno's paradox. Even an instantaneous photograph records the traces of movement, a sequence of events, however brief. But the idea of the *punctum temporis* is not only an absurdity logically, it is a worse absurdity psychologically.[76]

Trying to come to terms with this psychological absurdity, Gombrich on the one hand recalls St. Augustine's introspective account, in his *Confessions*, of memory and expectations somehow both being there in the consciousness of the present, and, on the other hand, the modern finding that "our impressions remain available for a brief span of time, the time that is known as the memory span or the specious

---

74 *Ibid.*, pp. 293 f.
75 *Ibid.*, pp. 296 and 301.
76 *Ibid.*, p. 297.

present", combining the latter finding with the notion of working memory.[77] He concludes that "the instant of which the theoreticians speak, the moment when time stands still, is an illicit extrapolation, despite the specious plausibility which the snapshot has given to this old idea".[78]

Time does not stand still when we look at a picture. We build the picture up in time, Gombrich writes, and hold "the bits and pieces we scan in readiness till they fall into places as an imaginable object or event"; we scan "backward and forward in time and space". And "we cannot estimate the passage of time in a picture", Gombrich stresses, "without interpreting the event represented."[79] To illustrate this latter point, he comments on some features in the iconography of the Presentation of the Virgin.

*Figure 4:* Titian, *Presentation of the Virgin.*
Compare Gombrich, *"Moment and Movement in Art"*

He refers, among others, to Titian's painting (Figure 4), with bystanders, in the picture, not looking at the scene itself but at each other, and with the large distance

---

77   *Ibid.*, p. 299. Gombrich uses the term "immediate memory", and in connection with the term "specious present" does not explicitly refer to William James, but there can be no misunderstanding as to what he is talking about.
78   *Ibid.*, p. 303.
79   *Ibid.*, p. 302.

the Virgin has to traverse from her family to the waiting priest, all of which extend the time span; and to Tintoretto's work (Figure 5), with "the steep curve of the steps" introducing yet another dynamic – temporal – effect.[80]

*Figure 5:*     *Tintoretto, Presentation of the Virgin.*
                 *Compare Gombrich, "Moment and Movement in Art"*

"If perception both of the visible world and of images were not a process in time", Gombrich writes by way of conclusion, "and a rather slow and complex process at that, static images could not arouse in us the memories and anticipations of movement."[81] A fitting formula to sum up his theory on image and time.

---

80    *Ibid.*, pp. 303 f.
81    *Ibid.*, pp. 305 f.

# 4. Image and Metaphor in the Philosophy of Wittgenstein

## Image-Blindness: A Prologue to Wittgenstein Scholarship

Five or so decades after the publication of the *Philosophical Investigations*, the passage most often cited when it comes to characterizing the later Wittgenstein's view of images in thought and communication is still § 115, regularly quoted together with § 116, making up the lines: "A *picture* held us captive. And we could not get outside it, for it lay in our language and language seemed to repeat it to us inexorably. – When philosophers use a word – 'knowledge', 'being', 'object', 'I', 'proposition', 'name' – and try to grasp the *essence* of the thing, one must always ask oneself: is the word ever actually used in this way in the language-game which is its original home? – What *we* do is to bring words back from their metaphysical to their everyday use."[1]

The 2004 volume *Wittgenstein's Lasting Significance*[2] is as good an example as any. It has much to say about the early picture theory of language, but practically nothing about the later Wittgenstein's philosophy of pictures, while recurrently using the phrase "being in the grip of a picture". To talk about pictures, it appears here, is to talk about words. Or take Anja Weiberg's paper from the same year, "'Ein Bild hielt uns gefangen': Die Kraft der Metapher",[3] where the author understands "picture" to mean, almost invariably, *figure of speech*. Her one notable exception is a brief reference to the phenomenon of seeing-as, a phenomenon Weiberg however immediately interprets as being grounded in linguistic, rather than in extra-linguistic, experience.[4] Now it is of course indeed the case that

---

1 Ludwig Wittgenstein, *Philosophical Investigations*, translated by G. E. M. Anscombe, second edition, Oxford: Basil Blackwell, 1958.
2 *Wittgenstein's Lasting Significance*, edited by Max Kölbel and Bernhard Weiss, London: Routledge, 2004.
3 Anja Weiberg, "'Ein Bild hielt uns gefangen': Die Kraft der Metapher", in *Wittgenstein und die Metapher*, edited by Ulrich Arnswald, Jens Kertscher and Matthias Kroß, Berlin: Parerga, 2004, pp. 115–135.
4 Weiberg, "'Ein Bild…'", p. 128.

what Wittgenstein, in § 115, had in mind, is not a visual image, and everyday language clearly permits, for good reasons, the use of the word "picture" in the sense of "view", "idea", "notion". What Wittgenstein here is saying is that our major handed-down philosophical notions were originally suggested, and are again and again reinforced, by certain figures of speech. But it is not at all the case that the later Wittgenstein invariably referred to linguistic formulas when he spoke of pictures. On the contrary, he had developed, as I will attempt to indicate, a variety of interesting ideas on how pictures function and visual images convey meaning, and how the verbal and the pictorial differ, and hang together – even if he did not succeed in synthesizing those ideas into a unified whole. Nor did he succeed in describing the kind of everyday usage alluded to in § 116, the usage he thought constituted the original paradigm metaphysical language ought to be led back to. "There is no trouble at all", Wittgenstein said in 1935, "with primitive languages about concrete objects. ... A substantive in language is used primarily for a physical body, and a verb for the movement of such a body. This is the simplest application of language, and this fact is immensely important. When we have difficulty with the grammar of our language we take certain primitive schemas and try to give them wider application than is possible."[5] Wittgenstein surely must have realized, but he appears to have been unable to come to terms with, the fact that everyday language has never been restricted to the kind of primitive languages he here invokes. Everyday language never was, and cannot be, devoid of metaphors.

Scholarship on Wittgenstein's philosophy of images does exist, but it does not amount to a continuous history; rather, it consists of a series of isolated attempts. When in 2000–2001 I had put together my first papers on the topic,[6] the awareness I had of those attempts was far from complete. The studies I referred to were writings by

---

5   "We might say", the passage continues, "that it is the whole of philosophy to realize that there is no more difficulty about time than there is about this chair", *Wittgenstein's Lectures: Cambridge, 1932–1935*, edited by Alice Ambrose, Oxford: Basil Blackwell, 1979, p. 119.

6   Kristóf Nyíri, "The Picture Theory of Reason" (2000), in *Rationality and Irrationality*, edited by Berit Brogaard and Barry Smith, Wien: öbv-hpt, 2001, pp. 242–266; Kristóf Nyíri, "Pictures as Instruments in the Philosophy of Wittgenstein" (2001), in *Wittgenstein and the Future of Philosophy: A Reassessment after 50 Years*, edited by Rudolf Haller and Klaus Puhl, Wien: öbv-hpt, 2002, pp. 328–336; Kristóf Nyíri, "Wittgenstein's Philosophy of Pictures" (2001), in *Wittgenstein: The Philosopher and his Works*, edited by Alois Pichler and Simo Säätelä (Working Papers from the Wittgenstein Archives at the University of Bergen, no. 17, 2005), pp. 281–312 (reprinted: Frankfurt/M.: ontos verlag 2006, pp. 322–353).

Gombrich,[7] Wollheim,[8] Kenny,[9] Genova,[10] Mitchell,[11] Roser.[12] Today I would add to that narrative of Wittgenstein research, from the early 1960s to the late 1990s, the names Aldrich,[13] Kjørup,[14] Blich,[15] Scholz,[16] Biggs,[17] Boehm,[18] and, with qualifications, Lüdeking.[19] Let me here present the narrative in a nutshell.

---

7   Ernst H. Gombrich, *Art and Illusion: A Study in the Psychology of Pictorial Representation*, London: Phaidon Press, 1960.
8   Richard Wollheim, *Art and Its Objects: An Introduction to Aesthetics*, New York: Harper & Row, 1968.
9   Anthony Kenny, *Wittgenstein*, Harmondsworth, Middlesex: Penguin Books, 1973.
10  Judith Genova, "Wittgenstein on Thinking: Words or Pictures?", in *Philosophy and the Cognitive Sciences*, edited by Roberto Casati and Graham White, Kirchberg am Wechsel: ÖLWG, 1993, pp. 163–167. See also her *Wittgenstein: A Way of Seeing*, London: Routledge, 1995.
11  W. J. T. Mitchell, *Picture Theory*, Chicago: The University of Chicago Press, 1994.
12  Andreas Roser, "Gibt es autonome Bilder? Bemerkungen zum grafischen Werk Otto Neuraths und Ludwig Wittgensteins", *Grazer Philosophische Studien* 52 (1996/97), pp. 9–43.
13  Virgil C. Aldrich, "Pictorial Meaning, Picture-Thinking, and Wittgenstein's Theory of Aspects", *Mind* 67 (1958), pp. 70–79.
14  Søren Kjørup, "Wittgenstein and the Philosophy of Pictorial Languages" (1980), in *Wittgenstein – Aesthetics and Transcendental Philosophy*, edited by Kjell S. Johannessen and Tore Nordenstam, Vienna: Hölder-Pichler-Tempsky, 1981, pp. 159–173.
15  Baruch Blich, "'Natural Kinds' As a Kind of 'Family Resemblance'", in *Philosophy of Law, Politics and Society*, edited by Ota Weinberger, Peter Kollee and Alfred Schramm (Proceedings of the 12th International Wittgenstein Symposium, 1987, Vienna: Hölder-Pichler-Tempsky, 1988), pp. 284–289.
16  Oliver R. Scholz, *Bild, Darstellung, Zeichen: Philosophische Theorien bildlicher Darstellung*, revised second edition, Frankfurt/M.: Klostermann, 2004, first edition published in 1991.
17  Michael A. R. Biggs, "A Source Catalogue of the Published Diagrams", in *Wittgenstein: Two Source Catalogues and a Bibliography*, edited by Michael Biggs and Alois Pichler (Working Papers from the Wittgenstein Archives at the University of Bergen, no. 7, 1993), pp. 91–143; Michael A. R. Biggs, *The Illustrated Wittgenstein: A Study of the Diagrams in Wittgenstein's Published Works*, PhD thesis, University of Reading, UK, 1994; Michael A. R. Biggs, "Graphical Problems in Wittgenstein's Nachlaß", in *Culture and Value: Philosophy and the Cultural Sciences*, edited by Kjell S. Johannessen and Tore Nordenstam, Kirchberg am Wechsel: ÖLWG, 1995, pp. 751–761.
18  Gottfried Boehm, "Die Wiederkehr der Bilder", in *Was ist ein Bild?*, edited by Gottfried Boehm, München: Wilhelm Fink Verlag, 1994, pp. 11–38.
19  Karlheinz Lüdeking, "Picture-Theory of Language and Language-Theory of Pictures", in *Wittgenstein – A Re-evaluation*, edited by Rudolf Haller and Johannes Brandl (Proceedings of the 14th International Wittgenstein Symposium, 1989, vol. III, Vienna: Hölder-Pichler-Tempsky, 1990), pp. 312–316.

Aldrich begins his 1958 paper by citing passages from Part I of the *Philosophical Investigations* where Wittgenstein uses the word "picture" primarily in the sense of "view", "idea", and only by implication in a visual sense. But he also refers to § 295, noting that Wittgenstein makes "the subtle point" that "the picture as an image" can be *evoked* by an expression,[20] and later in the paper[21] comes to Part II, sect. xi, mentioning the duck–rabbit drawing, and discussing the notions of "noticing an aspect", "picture-object", and "seeing something as something". However, Aldrich does not in the event provide even a rudimentary analysis of Wittgenstein's views on images, he does not explain what he means by his opening sentence "Wittgenstein has a theory of pictorial meaning and picture-thinking", and it is not by chance that the paper in fact exerted more influence on the topic of Wittgenstein and metaphor[22] than on the topic of Wittgenstein and images.[23] By contrast, Ernst Gombrich's reference to the duck–rabbit "trick drawing" and to its occurrence in the *Philosophical Investigations*, at the beginning of the introductory chapter of his seminal 1960 *Art and Illusion*,[24] could well have alerted scholarship to the fact that the later Wittgenstein had something important to say on the problem of pictorial meaning. But Gombrich, apparently, failed to have any impact on Wittgenstein scholarship, as did also, strangely, Richard Wollheim's 1968 *Art and Its Objects*, in which the author explicitly exploited and elaborated the Wittgensteinian notion of "seeing as".

Nor did any breakthrough come with Anthony Kenny's 1973 book *Wittgenstein*, in which the author, taking issue with the image of the "two Wittgensteins", stressed that the early "picture theory needs supplementing", rather than to be shown as false: the later "theory of meaning as use is a complement rather than

---

20 Aldrich, "Pictorial Meaning", p. 71.
21 *Ibid.*, pp. 73 f.
22 As evidenced by Marcus B. Hester, *The Meaning of Poetic Metaphor: An Analysis in the Light of Wittgenstein's Claim that Meaning Is Use*, The Hague: Mouton, 1967, and Rüdiger Zill, "Der Vertrakt des Zeichners: Wittgensteins Denken im Kontext der Metapherntheorie", in *Wittgenstein und die Metapher* (cf. note 3 above). Although the term "metaphor" does not actually occur in Aldrich's paper, it appropriately came to be reprinted in the volume *Essays on Metaphor*, edited by Warren A. Shibles, Whitewater, WI: Language Press, 1972, pp. 93–103.
23 But see the brief reference to Aldrich in Emmanuel Alloa, "Seeing-As, Seeing-In, Seeing-With: Looking Through Images", in *Image and Imaging in Philosophy*, edited by Elisabeth Nemeth, Richard Heinrich and Wolfram Pichler (33rd International Wittgenstein Symposium, Kirchberg am Wechsel: ALWS, 2010), p. 14.
24 Cf. note 7 above.

a rival to the picture theory".[25] However, Kenny's choice of the word "complement" did not mean that he had as it were detected, and found worth considering, a theory of pictures in Wittgenstein's later philosophy. In fact, Kenny in that book had absolutely nothing to say about the later Wittgenstein's views on pictures or images.[26] Almost another decade passed before there appeared the first study that actually had Wittgenstein's attempts at a theory of images as its subject: Søren Kjørup's "Wittgenstein and the Philosophy of Pictorial Languages", a talk given in 1980.[27] "Pictures", wrote Kjørup,

> always played an important role in the philosophical thought of Ludwig Wittgenstein. ... Wittgenstein never went so far as to formulate an explicit philosophy of pictures or philosophy of pictorial languages in its own right. ... But from his many asides on pictures and his many examples drawn from our use of and experience with pictures one does get a rather clear impression of his implicit conception of pictorial languages. ... And at certain points he even discusses pictures so straightforwardly and extensively that we come very close to an explicit theory.[28]

In his paper, Kjørup gives serious consideration to Wittgenstein's attempts, in *Philosophical Investigation*, Part II, sect. xi, to come to terms with the fact that pictures actually *depict*, that they represent by natural resemblance. Wittgenstein, as Kjørup puts it, does not deny in the *Philosophical Investigations* "that there is a connection between pictorial objects and real ones"; on the contrary, he asserts that towards, say, a "picture-face" one in some respects stands as one does towards a human face. "'I can study its expression, can react to it as to the expression of the human face. A child can talk to picture-men or picture-animals, can treat them

---

25   Kenny, *Wittgenstein*, p. 226.
26   In the same year that his book on Wittgenstein was published, Kenny had finished his translation of the so-called *Philosophical Grammar*, a mis-edition by Rush Rhees, as we today know, of Wittgenstein's TS 213 (the "Big Typescript"). One might hypothesize that Kenny here was overly impressed with Wittgenstein's formula, "Anything can be a picture of anything, if we extend the concept of picture sufficiently." (Ludwig Wittgenstein, *Philosophical Grammar*, translated by Anthony Kenny, Berkeley: University of California Press, 1974, p. 163.) Wittgenstein himself did not adhere to this formula for much longer; 1936 is the year it surfaces for the last time in his manuscripts. Kenny however is still in a way influenced by it as late as in 1993, in his book on Aquinas. As he there puts it: "there is good reason to believe that what makes an image of X an image of X is *never* its resemblance to X" (Anthony Kenny, *Aquinas on Mind*, London: Routledge, 1993, p. 99).
27   Cf. note 14 above.
28   Kjørup, "Wittgenstein...", p. 159.

as it treats dolls.'"[29] Wittgenstein, Kjørup points out, "here writes about our very direct and live relation to pictures: 'When I see the picture of a galloping horse – do I merely *know* that this is the kind of movement meant? Is it superstition to think I *see* the horse galloping in the picture?'" Wittgenstein in fact "stresses the difference between really experiencing a picture and just 'reading' it, as we might say: 'If you see the drawing as such-and-such an animal, what I expect from you will be pretty different from what I expect when you merely know what it is meant to be.'"[30] However, after having given due scrutiny to these remarks by Wittgenstein, Kjørup deems them to be misguided. By contrast, he embraces the Wittgensteinian approach according to which as a "point of departure for theorizing on pictures one should not take 'idle' pictures, but pictures in use".[31] The philosopher of images whose approach is in accordance with what the later Wittgenstein actually was up to, stresses Kjørup, is Nelson Goodman;[32] and what the later Wittgenstein was actually up to was the elaboration of a *use-theory of pictures*. These are ideas which today dominate the field.[33]

The first one to formulate an alternative set of ideas appears to have been Baruch Blich, in his 1987 Kirchberg talk "'Natural Kinds' As a Kind of 'Family Resemblance'".[34] Blich sets the tone of his argument by referring at quite some length to William Ivins' book *Prints and Visual Communication*,[35] in which the author points to the crucial role of pictorial representation in scientific argument and exposition, and to the verbal bias philosophy has suffered from for millennia, not possessing, and consequently not reflecting on, visual instruments. Blich underlines the fact that the later Wittgenstein, in the course of his philosophical expositions, not only exploits images, but that his "use of pictorial representation is far

---

29  I have quoted this remark by Wittgenstein in chapter 1 above, towards the end of the section "Meeting Rorty".
30  Kjørup, "Wittgenstein...", p. 168.
31  *Ibid.*, p. 171.
32  *Ibid.*, pp. 167 f. and 172.
33  Kjørup actually exerted a real influence here. Scholz, the leading German proponent of attributing a use-theory of pictures to Wittgenstein, refers in his *Bild, Darstellung, Zeichen* (cf. note 16 above) both to Kjørup's "George Inness and the Battle at Hastings, or Doing Things With Pictures", *The Monist* 58 (April 1974), pp. 216–235, and to his "Pictorial Speech Acts", *Erkenntnis* 12 (1978), pp. 55–71, though without mentioning his "Wittgenstein and the Philosophy of Pictorial Languages".
34  Cf. note 15 above.
35  William M. Ivins, Jr., *Prints and Visual Communication*, Cambridge, MA: Harvard University Press, 1953.

more than an illustration, and it is well embedded in his philosophical approach".[36] As Blich sees the matter, Wittgenstein's notion of family resemblance can help us to understand the way in which a depicted object and the picture depicting it can *resemble* each other:

> Language games and family resemblance ... play ... an important role, because only with the help of such understanding of language are we able not only to create a given context for identifying vague elements of pictures, but by expanding the language game we stretch our reality to include new things. ... Unless we could extend our language and apply words to pictures, one would not be able to grasp their relevance for reality, and this is true of simple pictures as well as of sophisticated pictures such as caricatures, impressionist paintings, cubist paintings etc.

Here Blich adds a momentous observation, remarking that Wittgenstein's "idea of language games and their place in constituting new meanings, new concepts etc. in a given language, can account for new and unconventional generalizations. Practically it means that a prediction or a generalization ... can from now on be expressed even by metaphorical expressions, similes and the like, not to mention pictures as such."[37]

Blich's 1987 talk, crucially important though it was, remained without impact.[38] Two years later Karlheinz Lüdeking gave a paper at Kirchberg,[39] in which he told about seeing striking parallels between Wittgenstein on the one hand, and René Magritte, "*the* one painter of classical modernity with a conspicuously Wittgensteinian attitude", on the other.[40] Lüdeking discussed Magritte's "Les mots et les images", reproducing and analyzing a fair number of the words-cum-images graphics from that little treatise (but *no* images from Wittgenstein's work), presenting Magritte as foreshadowing Goodman, and suggesting, practically, that the later Wittgenstein's views on pictorial meaning were quite similar to those of Goodman. Wittgenstein, writes Lüdeking, does already in his discussion of "the picture of the two fencers"[41] clearly distinguish "what the picture shows from what it stands for. And what the picture stands for, he indicates, is not determined by its own structure but by our use of the picture. ... In a modified form we find

---

36   Blich, "'Natural Kinds'", p. 285.
37   *Ibid.*, p. 288.
38   The only reference to the talk I am aware of is one by Dieter Mersch, in his "Wittgensteins Bilddenken", *Deutsche Zeitschrift für Philosophie* 54 (December 2006), p. 939, note 2.
39   Cf. note 19 above.
40   Lüdeking, "Picture-Theory...", p. 312.
41   Lüdeking here refers to Wittgenstein's notebook entry of 29. 9. 1914.

the same thought in a much later remark about a picture of a different fighting sport in the footnote to paragraph 22 of the *Philosophical Investigations*. The picture shows a boxer, but what it represents, and even that it represents anything at all, can only be inferred from our use of it."[42] Lüdekind has nothing to say on Wittgenstein's analyses of how pictures can have a direct, unmediated impact on us, as shown in particular in the *Blue and Brown Books*,[43] or, for that matter, in sect. xi, Part II, of the *Philosophical Investigations*. Similarly, Oliver Scholz, who in his – on the whole extremely rewarding – 1991 book *Bild, Darstellung, Zeichen*,[44] although referring extensively to passages in the writings of the later Wittgenstein that can very well suggest a different conclusion, unequivocally attributes to him a use-theory of pictures.

Subsequently to Blich's 1987 talk, the possibility of a major breakthrough arrived, again, with Judith Genova's 1993 Kirchberg paper, "Wittgenstein on Thinking: Words or Pictures?"[45]. As she there summed up the cognitive psychological background of her interpretation of Wittgenstein: "From an evolutionary perspective perhaps pictures represent an older form of thinking, one surpassed but never eliminated by words. ... Whatever the history, neither language nor thinking can do without their supplement of pictures. To the extent that we think in language, we think in pictures."[46] And this is how she began her talk: "contemporary epistemologists take words and pictures to be opposites. ... most would concur that thinking is discursive, not pictorial. ... Is thinking visual or verbal? – Wittgenstein's radical response is neither or either... In one sense, thinking is neither picturing nor speaking, but something else again. ... In another sense, however, thinking is *either* picturing or speaking. There is a family resemblance between the activities allowing for an exchange between them."[47] Words, Genova said, necessarily engender pictures. As she put it: "Pictures suffuse the speaking process. ... they make meaning possible by wedding the abstract word to a sensory embodiment."[48]

---

42    Lüdeking, *op. cit.*, p. 316.
43    Ludwig Wittgenstein, *Preliminary Studies for the "Philosophical Investigations": Generally Known as the Blue and Brown Books*, edited by Rush Rhees (1958), Oxford: Basil Blackwell, 1964, see esp. pp. 36 f., 105, 125, 162–174.
44    Cf. note 16 above.
45    Cf. note 10 above.
46    Genova, "Wittgenstein on Thinking", p. 166.
47    *Ibid.*, p. 163.
48    *Ibid.*, p. 164.

Just as with the earlier talk by Blich, Genova's 1993 paper, too, remained without echo.[49] In the same year, the Bergen Wittgenstein Archives published the milestone compilation by Michael Biggs, "A Source Catalogue of the Published Diagrams",[50] a work that for the first time called attention to the extent to which Wittgenstein made use of "non-textual material".[51] The catalogue identified 479 "graphic elements" in the works of Wittgenstein printed to that date, with Biggs' 1995 paper "Graphical Problems in Wittgenstein's Nachlaß" assessing the overall figure of published and unpublished graphic elements at 2500. By the mid-1990s a general awareness of the later Wittgenstein as a philosopher of not just the verbal but also of the visual, could well have emerged. But as a matter of fact it did not. An influential author, W. J. T. Mitchell, in his 1994 book *Picture Theory*, still spoke of "Wittgenstein's iconophobia and the general anxiety of linguistic philosophy about visual representation" as being "a sure sign that a pictorial turn is taking place", referring to "the apparent paradox of a philosophical career that began with a 'picture theory' of meaning and ended with the appearance of a kind of iconoclasm, a critique of imagery that led [Wittgenstein] to renounce his earlier pictorialism".[52] By contrast, Gottfried Boehm, in his seminal essay of the same year, "Die Wiederkehr der Bilder",[53] sees what he calls the "iconic turn" as actually brought about by the work of the later Wittgenstein. It was Wittgenstein, Boehm stresses, who, by detecting the pictorial hidden in the verbal, ultimately led the way from the linguistic turn to an iconic turn.[54] Boehm's crucial move is to spell out the fact that Wittgenstein's notion of family resemblances has an inevitably visual connotation: resemblances meet the eye, rather than speaking to abstract reason.[55]

---

49  Nor did her 1995 book *Wittgenstein: A Way of Seeing* receive much attention. I was relieved to register that, at the 2010 Kirchberg Wittgenstein Symposium, Marianne Richter repeatedly referred to Genova's book in her talk "Methodologische Aspekte des Bildgebrauchs bei Wittgenstein", in *Image and Imaging in Philosophy* (cf. note 23 above), pp. 271–273.
50  Cf. note 17 above.
51  Biggs' doctoral dissertation *The Illustrated Wittgenstein*, a work replete with novel ideas and significant bibliographical references, remains sadly unknown to the scholarly community.
52  W. J. T. Mitchell, *Picture Theory*, Chicago: The University of Chicago Press, 1994, pp. 12 f.
53  Cf. note 18 above.
54  "Wittgensteins Theorie bedeutet in der Geschichte der 'ikonischen Wendung' einen vorläufigen Endpunkt und insofern einen Durchbruch, als es die Befragung der Sprache war, welche der ihr innewohnenden Bildpotenz Nachdruck verschaffte, den linguistic turn in einen iconic turn überleitete" (Boehm, "Die Wiederkehr der Bilder", p. 14).
55  "Ähnlichkeiten stimulieren eine vergleichende Wahrnehmung, sie appellieren stärker ans Auge, als an den abstrakten Verstand" (*ibid.*).

Coming to the end of this preliminary narrative, let me refer to Andreas Roser's 1996 paper "Are There Autonomous Pictures? Remarks on the Graphic Work of Otto Neurath and Ludwig Wittgenstein",[56] a paper that was material to the awakening of my own interest in Wittgenstein's later philosophy of images. Wittgenstein's method of explaining philosophical points with the help of diagrams, Roser stresses, would have made no sense if he had really adhered to the position that images do not have an unequivocal meaning unless interpreted verbally. Roser's argument is that one could not speak of *different applications of the same picture* if one did not distinguish between the picture and its application. Pictorial meaning is of course not independent of our *use* of pictures. But nor is it independent of the fundamental equivalences between the structure of the picture on the one hand, and the structure of what it depicts, on the other.

## Wittgenstein's Philosophy of Pictures

What I had been attempting to show in my 2000–2001 papers on Wittgenstein's views on the role of images was, precisely, that those views included an awareness of pictures as *natural carriers of meaning* – the perspective Kjørup entertained way back in 1980 and then rejected. I referred, among many other passages in Wittgenstein's published writings, to one in the *Blue Book* where Wittgenstein calls attention to the possibility of "a picture which we don't interpret in order to understand it, but which we understand without interpreting it". There are, he writes, "pictures of which we should say that we interpret them, that is, translate them into a different kind of picture, in order to understand them; and pictures of which we should say that we understand them immediately, without any further interpretation."[57] I dwelled at length on some crucial passages in the *Brown Book* where Wittgenstein, touching on the issue of facial expressions, asks us to "contemplate the expression of a face primitively drawn in this way"[58]:

---

56 Cf. note 12 above.
57 Wittgenstein, *The Blue Book*, p. 36.
58 Wittgenstein, *The Brown Book*, p. 162.

We should let this face,[59] Wittgenstein continues, "produce an impression" on us. We will then say: "Surely I don't see mere dashes. I see a face with a *particular* expression." And the point Wittgenstein makes here is that we cannot actually explain *what* this particular expression consists in. As he puts it: "'Words can't exactly describe it', one sometimes says. ... It is as though we could say: 'This face has a particular expression: namely this' (pointing to something). But if I had to point to anything in this place it would have to be the drawing I am looking at." One has an experience here, Wittgenstein implies, which cannot be conveyed by *words*; although it *can* be conveyed by pointing to a drawing. It appears our system of communication is incomplete, unless *pictures* play a part in it.[60] Wittgenstein then goes on to describe two other cases where we would insist that we do not see "mere strokes" or "mere dashes". First, when we say "This is a *face*, and not mere strokes", distinguishing, for instance,

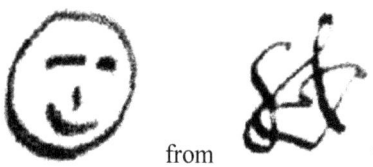

Secondly, the case of picture puzzles, when for instance "what at first sight appears as 'mere dashes' later appears as a face. We say in such cases: 'Now I see it as a face'." Wittgenstein stresses that this "seeing it *as a face*" does not indicate any delusions; rather, it "must be compared with seeing this drawing

either as a cube or as a plane figure consisting of a square and two rhombuses."[61] Some pages later, Wittgenstein experiments with the idea that instead of saying "I see this as a face" we should really say "I don't see this as a face, I see it like *this*". We should refrain from circumscribing verbally what we can simply *point to*.[62] What ought to have entered our verbal framework here, Wittgenstein implies,

---

59  The drawing is taken from the Bergen Electronic Edition, TS 310, p. 132. The same edition is the source of all subsequent graphics in the present chapter.
60  This implication will be somewhat later explicitly spelled out by Wittgenstein, cf. *The Brown Book*, p. 174.
61  Wittgenstein, *The Brown Book*, p. 163.
62  Wittgenstein, *The Brown Book*, p. 170.

is a non-verbal, *pictorial*, sign. This is the conclusion towards which the train of thought in the *Brown Book* in fact leads. And we are now in a position to see that what Wittgenstein in the so-called Part II of the *Philosophical Investigations* did was to take up, again, this train of thought. When studied together with the *Brown Book*, Part II of the *Philosophical Investigations* goes a long way towards giving a picture of what Wittgenstein's philosophy of pictures might amount to.

As to Part I of the *Philosophical Investigations*, among the passages I particularly referred to was § 450, where Wittgenstein relates calling up the image of someone to *mimicking* the person's expression. Since to mimic is to evoke a resemblance, and since the ability to mimic is in critical respects more fundamental than the ability to speak, Wittgenstein here again implies that some kinds of visual representation can convey meaning without relying on verbal appendage. I also drew attention to the occurrence of the word "picture" in the 1945 preface of the *Investigations*: he has produced an *album*, Wittgenstein there writes, made up of ever new pictures of the same sites.[63] The word "picture" is a metaphor here; but the metaphor – entirely absent in the 1938 version of the preface – is quite elaborate, the author likening himself to a poor draughtsman, with references made to picture cuts and to observers of landscapes. Significantly, in MS 130, where on p. 22 the term "album" first makes its appearance, the passage in which it occurs is in fact immediately followed by an interesting sequence of pictures: drawings in connection with the seeing-as issue (Figures 1 and 2).

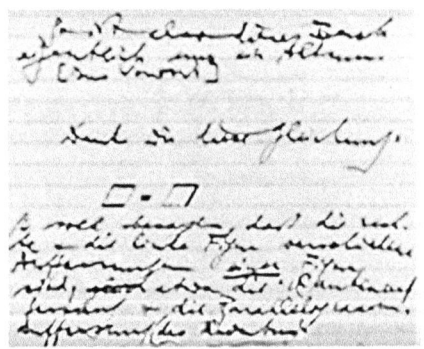

*Figure 1:*     *"verschiedene Auffassungen einer Figur"*
Source: *Wittgenstein's Nachlaß, MS 130, p. 22*

---

63 Wittgenstein, *Philosophical Investigations*, p. vii, third paragraph. The word "Bild" occurs twice in this paragraph, but in the English translation only the second occurrence is translated as "picture". The first occurrence, "immer neue Bilder entworfen", is rendered as "new sketches made".

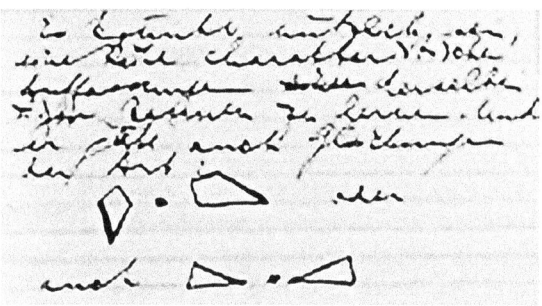

*Figure 2: "eine Reihe charakteristischer Auffassungen derselben Figur"*
*Source: Wittgenstein's Nachlaß, MS 130, p. 23*

This is one of the innumerable instances where a look at the *Nachlaß* context adds additional meaning to what Wittgenstein says in the printed version. Wittgenstein's published writings clearly offer a wealth of important ideas on the social function of pictures, on pictorial meaning, and on pictorial communication. These ideas however, as I argued in my 2000–2001 papers, do not add up to a unified philosophy of pictures. In fact, the later Wittgenstein at no stage of his thinking possessed such a unified philosophy. He had significant insights, but no clear views as to what his problems actually were, or what he was striving to achieve. Hence he often abandoned ideas his interpreters today might find promising; and many ideas never made it to the printed editions of his writings. No attempt at constructing a coherent philosophy of pictures out of his insights can then, I suggested, succeed without taking account of the entire *Nachlaß*; and I provided some examples of what working with the *Nachlaß* from this perspective might look like. One block of the Wittgensteinian corpus I should have covered, but did not, is the 1938 conversation notes edited by Cyril Barrett.[64] I do not here have the space to make up for that omission. But let me single out two truly extraordinary passages.

First, the passage "I remember walking in the street and saying: 'I am now walking exactly like Russell.' You might say it was a kinesthetic sensation. Very queer. – A person who imitates another's face doesn't do it before a

---

64 Ludwig Wittgenstein, *Lectures and Conversations on Aesthetics, Psychology and Religious Belief*, edited by Cyril Barrett, Berkeley: University of California Press, 1967. It was the title of the talk by Martin Kusch, "The Concept of Picture in Wittgenstein's 'Lectures on Religious Belief'", as announced in the preliminary program of the 2010 Kirchberg symposium, that alerted me to the relevance of this volume to a fuller understanding of Wittgenstein's philosophy of images.

mirror."[65] Imitating, creating a resemblance, Wittgenstein implies, is a primordially motor affair. It definitely has nothing to do with rules or conventions.

The second passage: the admission that, at the end of the day, *of course* picturing hinges on *likeness*. However else could it do its job? Wittgenstein talks about how portraits resemble: "If I give up the business of being like [as a criterion], I get into an awful mess, because anything may be his portrait, given a certain method of projection. ... If you're asked: 'How do you know it is a thought of such and such?' the thought that immediately comes to your mind is one of a shadow, a picture. You don't think of a causal relation. The kind of relation you think of is best expressed by 'picture', 'shadow', etc. – The word 'picture'", Wittgenstein here interjects, "is even quite all right – in many cases it is even in the most ordinary sense, a picture. You might translate my very words into a picture. – But the point is this, suppose you drew this [picture], how do I know it is my brother in America? Who says it is him – unless it is here ordinary similarity?"[66]

## Metaphor: The Stumbling-Block for Wittgenstein's Later Philosophy

A highly intriguing drawing by Wittgenstein is the one on p. 159 of MS 107, accompanying a remark jotted down on Nov. 10, 1929 (Figure 3):

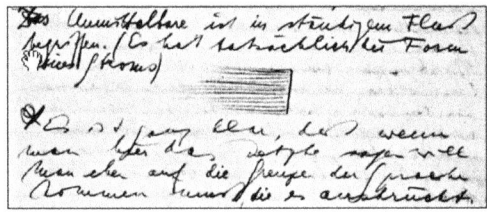

*Figure 3:   Drawing by Wittgenstein*
Source: *Wittgenstein's Nachlaß, MS 159, p. 159*

---

65  Wittgentein, *Lectures on Religious Belief*, p. 39. In Wittgenstein's manuscripts, two years later there occurs the formulation (MS 123, pp. 20 r/v): "Wissen wie jemand geht: es sich vorstellen können – aber auch: es nachmachen können. Muß man sichs vorstellen, um es nachzumachen? Und ist es nachmachen nicht ebenso stark, als es sich vorstellen?" By 1944 this becomes (MS 129, pp. 181 f., cf. the reference to PI § 450 above): "Wissen, wie jemand geht /ausschaut/: es sich vorstellen können – aber auch: es nachmachen /ahmen/ können. Muß man sich's vorstellen, um es nachzumachen? Und ist es nachmachen /ahmen/ nicht ebenso stark, als es sich vorstellen?"
66  Wittgentein, *Lectures on Religious Belief*, pp. 66 f.

"The immediate is in a constant flux [Fluß]. (It has in fact the form of a stream [Strom].) – It is quite clear that if one wants to say here the ultimate, one must thus come to the limit of the language which expresses it."[67] It is remarkable that Wittgenstein found himself able to at least *indicate* in a drawing something he implied one cannot *say*. However, what he at this transitory stage in his philosophy regarded as a *limit* of language, he soon came to see simply as its *lure*. As Wittgenstein put it in 1932, just because a sentence "sounds English", we take it to be "sensible": "Thus, for example, we talk of the flow of time and consider it sensible to talk of its flow, after the analogy of rivers. ... Discussion of 'the flow of time' shows how philosophical problems arise. Philosophical troubles are caused by not using language practically but by extending it... We form sentences and then wonder what they can mean."[68] Similarly in the *Brown Book*, where commenting on the question of the passage of time, Wittgenstein says: "It is clear that this question most easily arises if we are preoccupied with cases in which there are things flowing by us, – as logs of wood float down a river. ... We then use this situation as a simile for all happening in time and even embody the simile in our language, as when we say 'the present event passes by' (a log passes by), 'the future event is to come' (a log is to come). We talk about the flow of events; but also about the flow of time – the river on which the logs travel."[69]

Extending language by using analogies and similes – the problem Wittgenstein here is encircling is that of metaphor. The metaphor of the flow of time, he suggests,

---

67  With a single word changed, I am quoting this passage in the translation given by David Stern, in his pathbreaking "Heraclitus' and Wittgenstein's River Images", *The Monist* 74 (Oct. 1991), p. 588.
68  *Wittgenstein's Lectures* (cf. note 5 above), pp. 13 and 15.
69  Wittgenstein, *The Brown Book*, pp. 107 f. And a highly interesting *Nachlaß* passage, written roughly at the same time: "wenn uns beim Nachdenken über die Zeit das Bild des Vorüberfließens gefangen hält... Wie etwa, wenn wir an einem Fluß stehen auf dem Holz geflößt wird: die Stämme ziehen an uns vorüber; die, welche vorüber sind, sind alle rechts von uns, die noch kommen, sind links. ... Wir sprechen vom Lauf der Ereignisse, aber auch vom Laufe der Zeit, — des Flusses, auf dem die Stämme vorbeischwimmen. ('die Zeit ist da', 'die Zeit ist längst vorbei', 'es kommt die Zeit', etc., etc.) Und so kann mit dem Wort 'Zeit' das Bild eines ätherischen Flusses untrennbar verbunden sein, mit den Worten 'Vergangenheit' & 'Zukunft' das Bild von Gebieten, Ländern, aus deren einem die Ereignisse in das andre ziehen. Und doch können wir natürlich keinen solchen Strom finden & keine solchen Örter. Die Grammatik unserer Sprache läßt eben Fragen zu, zu denen es keine Antwort gibt. Und sie verleitet uns zu ihnen durch die Bildhaftigkeit des Ausdrucks. Eine Analogie hat unser Denken gefangen genommen & schleppt es unwiderstehlich mit sich fort" (MS 115, p. 172).

is a philosophically dangerous one, which we should avoid by keeping close to everyday – "practical" – language. However, Wittgenstein does not really seem to have made his case. He neither demonstrates that this metaphor invariably carries philosophical dangers, nor does he show that everyday language does not make spontaneous use of it.[70] And I think this instance is symptomatic of Wittgenstein's difficulty. As I have indicated at the beginning of the present chapter, the problem of metaphor, generally speaking, is one the later Wittgenstein has ultimately not been able to come to terms with.

The literature on Wittgenstein and metaphor is rich, but strangely discontinuous. From the point of view of the argument I pursue, the most important study here is Marcus Hester's 1967 book, *The Meaning of Poetic Metaphor: An Analysis in the Light of Wittgenstein's Claim that Meaning Is Use*.[71] Hester provides an historical overview of metaphor theory, beginning of course with Aristotle, and including, among many others, I. A. Richards, René Wellek, Austin Warren, Max Black, and Rom Harré. He observes that Wittgenstein's "remarks on metaphor are almost non-existent"[72] (this is an observation subsequent scholarship will invariably and repeatedly make, but it is actually wrong, if the entire *Nachlaß* is taken into consideration); registers (and exaggerates, as mainstream Wittgenstein scholarship to this day does) Wittgenstein's "attack on inner images"[73], arguing however that on this point Wittgenstein is wrong: language, and not just poetic language, does indeed rely on visual mental images;[74] and elaborates the position that building on

---

70 This is brilliantly argued by Walter Mesch. As he writes: "die Rede vom Vergehen, Verrinnen oder Verfliegen der Zeit [findet] bereits in der gewöhnlichen Sprache in vielen Varianten Verwendung… Dies dürfte kaum zu verstehen sein, wenn darin nicht irgendwelchen Erfahrungen Ausdruck verliehen wäre, die man im gewöhnlichen Leben machen kann. … Bei der Rede vom Fluss der Erscheinung oder Zeit scheint es sich nicht um eine falsche Verwendung der Sprache handeln zu können, die erst dann auftritt, wenn wir philosophieren. Wenn hier etwas zu kritisieren ist, scheint die Kritik auch auf den gewöhnlichen Sprachgebrauch bezogen werden zu müssen; und dies kann für jemanden, der seine Hauptaufgabe darin sieht, gegen die vermeintlichen Selbstverständlichkeiten des philosophischen Sprachgebrauchs auf den gewöhnlichen Sprachgebrauch zu verweisen, keineswegs unproblematisch sein. Es sieht so aus, als drohten sich Wittgensteins Einsichten gegen ihn selbst zu wenden." (Mesch, "Die Metaphern vom Vergehen und vom Fluss der Zeit: Überlegungen im Anschluss an eine Bemerkung Wittgensteins", in *Wittgenstein und die Metapher* [cf. note 3 above], pp. 273 and 277.)
71 Cf. note 22 above.
72 *The Meaning of Poetic Metaphor*, p. 31.
73 *Ibid.*, pp. 37 ff., cf. also p. 34.
74 *Ibid.*, pp. 69, 92, 96, and esp. pp. 133 ff.

the one hand on Wittgenstein's theory of meaning as use, and, on the other hand, on the insight that language use actually involves evoking images,[75] a theory of metaphor can be defended which vindicates the role of imagistic thinking,[76] but would seem to be unacceptable to Wittgenstein.[77]

Hester's suggestions, at the time he made them in the mid-1960s, must have sounded entirely outlandish to the philosophical community. My impression is that his outstanding book remained largely without influence. There was one notable exception: Paul Ricoeur, in his 1975 study *La métaphore vive*,[78] did indeed discuss, and to some extent even assimilate, the connection Hester had established between image and metaphor. And Ricoeur's work, of course, has been widely read and cited. His references to Hester, however, went unnoticed. Jerry Gill, in his book *Wittgenstein and Metaphor*,[79] does acknowledge Ricoeur, but is unaware of Hester's book. He stresses that although, clearly, "Wittgenstein had no explicit theory of metaphor", it "is just as clear … that his writings contain an implicit view of the nature and significance of metaphorical speech". And the background of this implicit view is that the notion of language, as put forth in the *Investigations*, "is congenial to the notion of metaphoric meaning by reason of its stress on the flexibility and functionality of linguistic phenomena. … Wittgenstein's *use* of metaphor embodies a view of metaphor as both primordial and cognitive". However, "the literature on the role of metaphor in Wittgenstein's *Philosophical Investigations* is virtually non-existent".[80] The primary significance of Wittgenstein's work for philosophy, Gill believes, "lies in his suggestion that at the most fundamental level philosophy is a metaphorical enterprise". But Gill also maintains, and he appears to sense no tension here, that "Wittgenstein relies most heavily upon the metaphoric mode, especially as it constitutes the heart of everyday speech, because it is at the practical level of existence that we are closest to the bedrock of our form of life".[81]

---

75 *Ibid.*, pp. 98 f., 113.
76 *Ibid.*, pp. 39, 111, 176 f.
77 *Ibid.*, pp. 23 f., 96, 108, 113, 191.
78 English translation: *The Rule of Metaphor: The Creation of Meaning in Language*, London: Routledge, 2003. Well before the English translation, a German one was published: *Die lebendige Metapher*, München: Wilhelm Fink, 1986.
79 Jerry H. Gill, *Wittgenstein and Metaphor* (1981), new and revised edition, Atlantic Highlands, NJ: Humanities Press, 1996. A first draft was Jerry H. Gill, "Wittgenstein and Metaphor", *Philosophy and Phenomenological Research* 40 (1979), pp. 272–284.
80 Gill, *Wittgenstein and Metaphor*, p. 82.
81 *Ibid.*, pp. 99, 108 f., 128 and 130.

While Gill was unaware of Hester, the authors of the 2004 volume *Wittgenstein und die Metapher*[82] are aware neither of Hester nor of Gill. As the editors claim in their introduction to the volume: although the topic of metaphor has been widely discussed in the analytic tradition, and although the use of metaphorical language in the texts of some leading twentieth-century philosophers has received detailed scholarly attention, nothing similar has been attempted with regard to the philosophy of Wittgenstein. Also, the editors point out that while there are numerous places in Wittgenstein's later writings which indeed raise, for his philosophy, the question of how the border between literal and non-literal linguistic usage should be conceived of, still, "one would search in vain for a theory of metaphor, or even the beginnings of such a theory, in his work".[83] One of the authors of the volume, Matthias Kroß, in his chapter "The Self-Evidence of Metaphor: Wittgenstein's Relaxation of a Problem in the Philosophy of Language", again remarks that Wittgenstein practically never voiced an explicit opinion on the issue of metaphor, while his deliberations on language-games and on concepts bearing family resemblances to each other, have a clear implication: it does not make sense, anymore, to speak of the "original", the "literal" application of a concept. Some pages later, Kroß comes to describe Wittgenstein's diagnosis of the ultimate source of philosophy's ever unsolvable problems: these problems arise out of a misapplication of language, out of a *carrying over* of some specific linguistic usage from one sphere of discourse to another.[84] But here, nearing the end of this chapter, I am compelled to interject a question: why should that carrying over count as a misapplication? Kroß as it were highlights a contradiction in Wittgenstein's later philosophy; but he does not seem to realize the fact. Two authors of the same volume who do see this contradiction are Walter Mesch, discussing the unconvincing way Wittgenstein deals with the flow of time simile,[85] and Rüdiger Zill,[86] referring to an early paper by Warren Shibles, in which the latter pointed out the discrepancies in Wittgenstein's attitude towards metaphor.[87]

Shibles sides with the view that language is primarily metaphorical. For Wittgenstein however, as Shibles writes, "whereas a language-game can

---

82   Cf. note 3 above.
83   "Eine Theorie der Metapher oder Ansätze dazu wird man in Wittgensteins Werk ... vergeblich suchen" (*Wittgenstein und die Metapher*, pp. 11 f.).
84   Kroß, "Die Selbstverständlichkeit der Metapher: Wittgensteins Entspannung eines sprachphilosophischen Problems", in *Wittgenstein und die Metapher*, pp. 31 f. and 34.
85   Cf. note 70 above.
86   Cf. note 22 above.
87   For Zill's reference to Shibles, see *Wittgenstein und die Metapher*, pp. 162 f.

change we must try to stick with the literal, original language-games we learned. ... Wittgenstein underplays the notion of metaphor and instead concentrates on getting language back into what he calls 'ordinary' language. This underplay of metaphor however only accords with his explicit statement. In actual practice, as we can see by Wittgenstein's style of presentation and argument, he is a master at gaining insight by the use of analogy, metaphor and striking juxtapositions."[88] There are two levels of tension here. At the surface level there is the tension between, on the one hand, Wittgenstein's not giving theoretical weight to metaphor, and, on the other, his exuberant use of it. At the more fundamental level, there is a straightforward contradiction between Wittgenstein's claim of the primordial literalness of everyday language, and his stress on the multiplicity and flexibility of language-games.[89] It is not at all the case that Wittgenstein was not occupied with the problem of metaphor. Especially MS 150 (1935–36), MS 152 (1936) and the later parts of MS 115 (1936) offer rich material on "literal meaning" (*eigentliche Bedeutung*) and "transposed meaning" (*übertragene Bedeutung*). In 1947, Wittgenstein jotted down a telling passage: "But it is surely important that ... worry can be described in such words as: 'the descent of a permanent cloud'. I have perhaps never stressed the importance of this paraphrasing enough. – Think of happiness portrayed through a face surrounded by light, by rays emanating from it."[90] Wittgenstein's problem was that he did not succeed in making his ideas on metaphor, and indeed his ideas on metaphor and images, converge with the main drift of TS 227 (the so-called "Part I" of the so-called "Philosophical Investigations"). It was this divergence, I believe, that prevented him from rounding out his later philosophy.

---

88  Warren A. Shibles, *Wittgenstein, Language and Philosophy*, Dubuque, IA: Kendall/Hunt, 1969, pp. 2 f.
89  This is the "incompatibility", I believe, Schulte ultimately hints at in his "Wittgenstein's Notion of Secondary Meaning and Davidson's Account of Metaphor – a Comparison", *Grazer Philosophische Studien* 36 (1989), p. 145.
90  "Es ist aber doch wichtig, daß ... man die Sorge mit den Worten beschreiben kann 'Ewiges Düstre steigt herunter'. Ich habe vielleicht die Wichtigkeit dieses Paraphrasierens nie genügend betont. – Man stelle die Freude dar durch ein lichtumflossenes Gesicht, durch Strahlen, die von ihm ausgehen." (MS 134, p. 52. The first part of this passage has been published as § 517 of *Zettel*, translated by G. E. M. Anscombe, second edition, Oxford: Basil Blackwell, 1981. Miss Anscombe renders "Sorge" as "care".)

# 5. Time As a Figure of Thought and As Reality

## Figures of Thought: A Preliminary Outline

Although the term "figure of thought" has come to be very much in vogue, it seems to lack any clear definition.[1] I am not attempting to provide such a definition here, but I do endeavour to offer a concise characterization: I conceive of figures of thought as *mediating* between *different dimensions* of experience, thinking, and communication; such as the motor, the visual, the verbal, and even the musical.

Let me, by way of introduction, present some figures of thought selected in this spirit. First, a metaphor. I believe any live metaphor would do, but, to make my point, I am choosing a specific metaphor you have possibly not yet encountered, a metaphor coined by cognitive scientist Allan Paivio. It is a metaphor designed to show that processing a metaphor relies on processing mental images. This is how it runs: "for the student of language and thought, metaphor is a solar eclipse". The meaning Paivio intended to convey is that in a metaphor, just as in an eclipse, something is obscured; but also, that both a metaphor and an eclipse enlighten while they obscure. Paivio has put this metaphor to test subjects, and found that in order to understand it they first, indeed, had to *visually imagine* the eclipsed sun.[2] To understand a live metaphor, then, means to move from words to images, and then back from images to words. A second figure of thought I am offering: a saying conveyed via a depiction. Some hundred such depictions can be found in Pieter Brueghel the Elder's 1559 painting *Netherlandish Proverbs*. Take the saying "Big fish eat little fish". In the painting, you can spot the tiny bit rendering this very saying in pictorial form. That bit is a figure of thought, it is the result of

---

1 Some main approaches to the topic, and the wide divergences they display, were impressively summarized in the October 2010 call for papers for the workshop "Was sind Denkfiguren? Figurationen unbegrifflichen Denkens in Metaphern, Diagrammen und Kritzeleien". The workshop took place on February 25–26, 2011, at the Freie Universität Berlin, where I had the privilege to read an early version of the present chapter.

2 See Allan Paivio – Mary Walsh, "Psychological Processes in Metaphor Comprehension and Memory", in Andrew Ortony (ed.), *Metaphor and Thought* (1979), rev. second edition, Cambridge: Cambridge University Press, 1993.

a movement from the verbal to the visual, and itself of course triggers a reverse movement from the pictorial to the verbal, namely the proverbial. The proverb is about powerful people or institutions defeating the less powerful ones; understanding the proverb involves forming mental images; and those mental images seem not only to picture the literal meaning of the proverb, but also to help one grasp its broader, idiomatic meaning. And a third figure of thought: a caricature. Almost any caricature would do, but let me choose the famous sequence Ernst Gombrich reprints in his *Art and Illusion*, a sequence published in a satirical paper in France in 1834, demonstrating how a portrait of Louis Philippe can be transformed into the picture of a pear, *poire* meaning "fathead" (Figure 1).[3] This is a pictorial metaphor with the meaning *the king is a fathead*; a figure of thought, leading from the verbal to the visual.

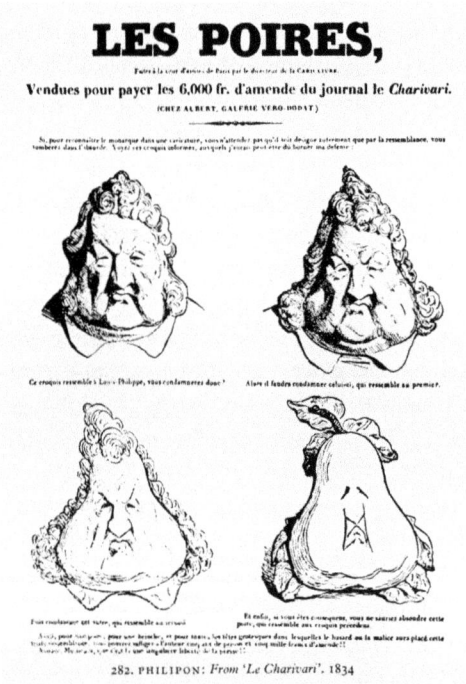

*Figure 1:*   *Caricature of Louis Philippe, by Philipon (1834)*

---

3   Ernst H. Gombrich, *Art and Illusion: A Study in the Psychology of Pictorial Representation* (1960), London: Phaidon Press, 1962, p. 291.

# Death, Music, and Time:
# Scribbles by Wittgenstein, Scribbles by Arnheim

To this day, the dominant image of Wittgenstein is that of a philosopher of language. Actually, as I am arguing throughout this book, he was a visual thinker: as much a philosopher of pictorial as of linguistic meaning, with diagrams and drawings abounding in his manuscripts and in the notes taken by his students. Let me first single out a scribble printed in the *Lectures and Conversations* edited by Barrett. Wittgenstein discusses the conditions under which someone can, or cannot, meaningfully speak about having an idea of something. The example introduced is the idea of death, with Wittgenstein insisting that for anyone's idea of death to be meaningful, the application of that idea must have public criteria. "If what he calls his 'idea of death' is to become relevant", Wittgenstein says, "it must become part of our game. – 'My idea of death is the separation of the soul from the body' – if we know what to do with these words. He can also say: 'I connect with the word "death" a certain picture – a woman lying in her bed' – that may or may not be of some interest. – If he connects" – and now comes a scribble (see Figure 2) – "with death, and this was his idea, this might be interesting psychologically."[4] To be sure, this scribble is not an established element of our language-game – hence the qualification "psychologically" – but it can certainly serve as a basis for making points about the idea in question. Our imagination, as Wittgenstein wrote, is a "complicated formation out of heterogeneous components – words, pictures" (this is a formula he again and again used);[5] scribbles distil such compounds into the purely visual, and in turn give rise to verbal formulae describing what we come to see.

Figure 2:   *"Death" – scribble by Wittgenstein*

Another of Wittgenstein's scribbles I will reprint here occurs in a brief memoir written by his student John King. King had a gramophone at his student room in

---

4   Ludwig Wittgenstein, *Lectures and Conversations on Aesthetics, Psychology and Religious Belief*, edited by Cyril Barrett, Berkeley: University of California Press, 1967, p. 69.
5   Cf. e.g. Ludwig Wittgenstein, *Philosophical Grammar*, translated by Anthony Kenny, Berkeley: University of California Press, 1974, p. 162.

Cambridge, and in the early 1930s Wittgenstein visited him several times to listen to music. "I once put on", King writes, "the second, third and fourth movements of Beethoven's Quartet in C sharp minor, *op*. 131. ... [Wittgenstein] was rapt in his attention and most excited at the end of the playing. He jumped up as if something had suddenly struck him and said, 'How easy it is to think that you understand what Beethoven is saying' (and here he seized a pencil and a piece of paper) 'how you think you have understood the projection' (and he drew two-thirds of a circle) 'and then suddenly' (and here he added a bulge) you realize that you haven't understood anything at all'."[6] (See Figure 3.)

*Figure 3:*    *"What Beethoven is saying" – scribble by Wittgenstein*

I doubt if this scribble, or pair of scribbles, actually expresses some important insight by Wittgenstein.[7] As you of course know, op. 131 occupies a special, keystone, place in Beethoven's oeuvre. It consists of seven movements to be played without a break, the fourth movement, the central one, being a set of seven variations on a simple theme. I assume it is a sound strategy to concentrate on this theme when trying to make sense of Wittgenstein's scribbles, and I submit that unless one takes them to allude to the fact that the basic theme is built up by an interplay of the two violins – a rather trivial allusion – they do not convey anything essential as regards the quartet in question. However, they convey the important fact that, although Wittgenstein was usually quite explicit on the dangers of attempting to describe the musical in non-musical terms,[8] still, on at least one occasion he could

---

6    John King, "Recollections of Wittgenstein", in Rush Rhees (ed.), *Recollections of Wittgenstein*, rev. ed., Oxford: Oxford University Press, 1984, pp. 69–70.

7    Thus I am not convinced, although very much impressed, by what Katrin Eggers in this connection suggests in her paper, "Form und Inhalt in der Musik – Wittgensteins Beitrag zu einem zentralen musikphilosophischen Problem", in *Image and Imaging in Philosophy, Science, and the Arts,* edited by Elisabeth Nemeth, Richard Heinrich and Wolfram Pichler, 33rd International Wittgenstein Symposium, Kirchberg am Wechsel: ALWS, 2010, pp. 74 f.

8    In the *Brown Book*, Wittgenstein talks about the "illusion" that "possesses us", when "repeating a tune to ourselves and letting it make its full impression on us, we say 'This tune says *something*', and it is as though I had to find *what* it says. And yet I know that it doesn't say anything such that I might express in words or pictures what

not but yield to the urge to express a musical impression visually; could not but yield to the urge, in his excited state, to turn to a figure of thought.

A third drawing by Wittgenstein I want to draw attention to here I have reprinted as Figure 3 in the previous chapter. It concerns the problem of the flow of consciousness and/or the flow of time, and it is not, strictly speaking, a scribble; it is, rather, a drawing representing a conventional metaphor in visual form. Indeed, viewed from Wittgenstein's perspective, it amounts to an attempt to *draw* something one cannot *say*. I believe that Wittgenstein's perspective is wrong, and I will come back to this drawing in the concluding section of the present chapter. Just now I will turn to a different set of scribbles having to do with time, scribbles published in Rudolf Arnheim's 1969 book *Visual Thinking*.

I have presented Arnheim, in the chapters above, as a pioneering central figure in the counter-attack on the linguistic turn in twentieth-century philosophy and psychology. As he puts it in his seminal 1969 book: "What makes language so valuable for thinking … cannot be thinking in words. It must be the help that words lend to thinking while it operates in a more appropriate medium, such as visual imagery."[9] This is, incidentally, the passage Lakoff quotes by way of introduction, when recounting, in a 2006 essay, the formative and then suppressed influence Arnheim's *Visual Thinking* had on him in the 1970s.[10] In his book, Arnheim dwells at length on the connection between abstract concepts, mental images, and drawings. "The prototype of drawings I have in mind", he writes, "are those diagrammatic scribbles drawn on the blackboard by teachers and lecturers in order to describe constellations of one kind or another – physical or social, psychological or purely logical."[11] In several experiments, Arnheim asked his students to produce spontaneous scribbles representing specific concepts. One group was instructed to draw *Past, Present, and Future*. Here I reproduce three of the drawings, the first two with explanations added by Arnheim (Figure 4).

---

it says. And if, recognizing this, I resign myself to saying 'It just expresses a musical thought', this would mean no more than saying 'It expresses itself'." (Ludwig Wittgenstein, *Preliminary Studies for the "Philosophical Investigations": Generally Known as the Blue and Brown Books*, edited by Rush Rhees [1958], Oxford: Basil Blackwell, 1964, p. 166.)

9    Rudolf Arnheim, *Visual Thinking*, Berkeley: University of California Press, pp. 231–32.

10   George Lakoff, "The Neuroscience of Form in Art", in Mark Turner (ed.), *The Artful Mind: Cognitive Science and the Riddle of Human Creativity*, Oxford University Press, 2006, p. 154.

11   Arnheim, *op. cit.*, p. 116.

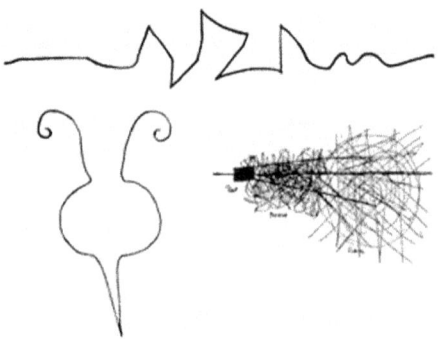

*Figure 4:* "Time" – *scribbles conveyed by Arnheim.
After Arnheim, Visual Thinking*

First, a continuous line. It indicates, Arnheim writes, "a straight and perhaps empty past, large and articulate shapes for the present, and some smaller and vaguer ones for the future. ... the whole of life is represented as an unbroken flow of time." Second, a pattern showing, as Arnheim interprets it, "gradual expansion, starting with the moment of birth". There is a break maintained "between past and present ..., but the largeness of the present is understood in part as a result of the preceding growth. The undirected roundness of the present interrupts the channeling of time, and yet this static situation in the middle of the drawing is ... traversed by a current of movement initiated in the past and carried further into the open future, as a river flows through a lake." Third, a structurally very different drawing, with the explanation given by the young draftsman himself: "The *past* is solid and complete, but still influences the present and the future. – The *present* is complex and not only a result of the past and leading to future, thus overlapping both, but is an entity in itself (black dot). – The *future* is least limited but influenced by both, past and present. – One line runs through for all have one common element – time."[12]

## Arnheim on Gestures and Scribbles

According to Arnheim, scribbles, far from being arbitrary drawings, express the very essentials of our thought processes. The argument he offers consists of two basic steps. In the first step, Arnheim relates line drawings to their "forerunners", namely

---

12   *Ibid.*, pp. 120 ff.

descriptive gestures. He points out that "the portrayal of an object by gesture rarely involves more than some one isolated quality or dimension… By the very nature of the medium of gesture, the representation is highly abstract. … Often a gesture is so striking because it singles out one feature relevant to the discourse." In the second step, Arnheim suggests that what a descriptive gesture pictures is not primarily a mental image, but rather the motor experience underlying that image. As he writes: "Gestures enact pushing and pulling, penetration and obstacle… the perceptual qualities of shape and motion are present in the very acts of thinking depicted by the gestures and are in fact the medium in which the thinking itself takes place. These perceptual qualities are not necessarily visual or only visual. In gestures, the kinesthetic experiences of pushing, pulling, advancing, obstructing, are likely to play an important part."[13] What Arnheim here says is, I believe, of paramount importance, since it implies not only that our verbal constructs – direct designations, idioms, metaphors – are meaningful because they convey mental images, but also that those images are given rise to by bodily, physical experiences, by our physical contact with reality. Scribbles depicting the flow of time are telling us something about what time really is. This goes very much beyond the position Lakoff and his school ever ventured to take.

## Image and Time in Conceptual Metaphor Theory

The locution "figure of thought" is not a household phrase in the Lakoff school. However, the idea of images mediating between words is of course quite central to conceptual metaphor theory.[14] The notion of *image schemas*, not yet present in the book *Metaphors We Live By*, but assuming an essential role by 1987 both in Johnson's *The Body in the Mind*[15] and in Lakoff's *Women, Fire, and Dangerous Things*,[16] is presented as an explicitly Kantian one,[17] linking perception and reason.[18] Image schemas are abstract visual/conceptual structures, not

---

13  *Ibid.*, pp. 117 f. I have already quoted this passage in chapter 1 above, in the subsection "The Visual and the Motor".
14  I am deeply indebted to Zoltán Kövecses for enlightening conversations on the topic of image and metaphor.
15  Mark Johnson, *The Body in the Mind: The Bodily Basis of Meaning, Imagination, and Reason*, Chicago: The University of Chicago Press, 1987. Compare my brief reference to Johnson and the notion of image schemas in the last section of chapter 2 above.
16  George Lakoff, *Women, Fire, and Dangerous Things: What Categories Reveal about the Mind*, Chicago: University of Chicago Press, 1987.
17  See *The Body in the Mind*, pp. 21 and 24, and *Women, Fire, and Dangerous Things*, p. 453.
18  See esp. *Women, Fire, and Dangerous Things*, p. 440.

to be confused, as Johnson and Lakoff again and again emphasize, with full-fledged mental images, actual mental *pictures*. But these latter images also play an increasingly important part in the Lakoff–Johnson approach. In *Metaphors We Live By*, the authors come to discuss what they call "two subcases" of the TIME PASSES US metaphor. In one case, they write, "we are moving and time is standing still; in the other, time is moving and we are standing still." These two metaphors, as they put it, "are not consistent (that is, they form no single image)", but they are nonetheless *coherent*, they "fit together".[19] The idea that metaphors can evoke visual images,[20] but that "a single consistent concrete image"[21] will not necessarily emerge when several "coherent but not consistent" metaphors overlap,[22] is a recurrent one in *Metaphors We Live By*; but no attempt is made here to establish a systematic connection between metaphor and imagery. By contrast, the topic of images very much comes to the fore in *Women, Fire, and Dangerous Things*, most notably in a discussion of what Lakoff calls *imageable idioms*, idioms relying on "conventional images". Lakoff provides an elaborate analysis of the idiom *to keep someone at arm's length*. "I have asked hundreds of people", he writes, "if they have an image associated with this idiom. Almost everyone does, and it is almost always the same image".[23] Then in the 1989 Lakoff–Turner volume *More than Cool Reason* there appears the notion of an "image metaphor". Quoting André Breton's lines "My wife ... Whose waist is an hourglass"[24], the authors refer to "a superimposition of the image of an hourglass onto the image of a woman's waist by virtue of their common shape. ... the metaphor is conceptual; it is not in the words themselves. ... the locus of the metaphor is [a] mental image."[25] The 1999 Lakoff–Johnson book *Philosophy in the Flesh* has a sub-chapter on "Metaphorical Idioms and Mental Imagery", describing a cognitive pattern where words evoke images that carry specific, conventional knowledge;[26] metaphorical idioms as

---

19  George Lakoff – Mark Johnson, *Metaphors We Live By*, Chicago: University of Chicago Press, 1980, p. 44. Compare also note 63 in chapter 2 above.
20  Cf. e.g. *Metaphors We Live By*, p. 168.
21  *Ibid.*, p. 105.
22  *Ibid.*, p. 94.
23  *Women, Fire, and Dangerous Things*, p. 447.
24  Translation by David Antin.
25  George Lakoff – Mark Turner, *More than Cool Reason: A Field Guide to Poetic Metaphor*, Chicago: The University of Chicago Press, 1989, p. 90, cf. p. 93.
26  George Lakoff – Mark Johnson, *Philosophy in the Flesh: The Embodied Mind and Its Challenge to Western Thought*, New York: Basic Books, 1999, pp. 67 ff.

imageable idioms are, no doubt, fundamental figures of thought. The book *Philosophy in the Flesh* is also where we encounter Lakoff and Johnson's most elaborate treatment of the philosophical problem of time. As they sum up the issue, "it is virtually impossible for us to conceptualize time without metaphor. ... Most of our understanding of time is a metaphorical version of our understanding of motion in space."[27] Hence "spatial metaphor for time" is part of our "cognitive unconscious" that structures, Lakoff and Johnson write, "not only the way we conceptualize the relationship between events and time but the very way we experience time".[28] However, the authors do not seem to have the courage of their convictions. They conclude that "we cannot take the common-sense understanding of time at face value from a cognitive perspective", and that the question "does time exist independent of minds?" should be rejected, rather than answered along the lines common-sense metaphors would suggest.[29] It is this conclusion I venture to take issue with.

## The Reality of Time

In their 1980 book, Lakoff and Johnson had already emphasized that the ultimate source of our fundamental metaphors are the experiences we have with physical objects, especially the experiences relating to our own bodies.[30] By 1987, their stress on the role of the kinesthetic, the motor, had become even more pronounced. Lakoff, in *Women, Fire, and Dangerous Things*, points out that "much of mental imagery is kinesthetic – that is, it is independent of sensory modality and concerns awareness of many aspects of functioning in space: orientation, motion, balance, shape judgments, etc.", and that the same holds even more for "image schemas, which are sufficiently general in character to be prime candidates for having a kinesthetic nature".[31] Johnson, in *The Body in the Mind*, defined image schemas as "recurring, dynamic pattern[s] of our perceptual interactions and motor programs",[32] offering, as an example, the COMPULSIVE FORCE schema, underlining that the concept "force" emerges from

---

27 *Ibid.*, p. 139. I have already quoted this passage in the present volume, cf. chapter 2, note 62.
28 *Ibid.*, p. 153.
29 *Ibid.*, pp. 154 and 167.
30 *Metaphors We Live By*, p. 25.
31 *Women, Fire, and Dangerous Things*, p. 446.
32 *The Body in the Mind*, p. xiv, cf. note 60 in chapter 2 above.

our *bodily experience* of force, from our encountering obstacles that exert force on us, from "the experience of being moved by external forces, such as wind, water, physical objects, and other people",[33] the experience that, say, "[w]hen a crowd starts pushing, you are moved ... by a force you seem unable to resist",[34] and from our experience that we too can exert force on, in some cases even penetrate through, the objects resisting us. I believe that our experience of *the passing of time*, too, amounts to an experiencing of some external force. We are all acquainted with what can without exaggeration be called the feeling of brute muscular tensions when struggling against time.[35] I suggest that a plausible metaphor to associate with the COMPULSIVE FORCE schema might be this one: THE PASSAGE OF TIME IS A PHYSICAL FORCE.

The passage of time means that the present becomes past, and that the future becomes present. However, the second part of this formula is misleading. As Broad used to point out, the future, strictly speaking, does not exist.[36] Instead of saying that the future becomes present, we should say that ever new presents come into being, or, still more precisely, that so far as the course of our own life is involved, ever new presents are created by us. Creating new presents is what struggling with time means. Metaphorically speaking, creating the future requires physical force.

Let us now come back to Wittgenstein's drawing of the flow of consciousness and/or the flow of time. Another of his remarks may help us interpret this drawing. "It is strange", he wrote in 1930, "that in ordinary life we are not troubled by the feeling that the phenomenon is slipping away from us, the constant flux of appearance, but only when we philosophize. ... The feeling we have is that the present disappears into the past without our being able to prevent it."[37] What we see in the drawing, then, is the present represented as a vertical line, with the horizontal lines, moving to the left, representing the present as changing into an ever more distant past. The later Wittgenstein became increasingly unhappy with the flow of time metaphor; he came to see it as an instance of the sickness of language that philosophy amounts to. I believe he was wrong. I very much agree with what Walter Mesch says: "talk about the passage of time is in no way restricted to philosophers ..., but is in many variants an element of everyday

---

33  *Ibid.*, p. 45.
34  *Ibid.*
35  Cf. my references to Münsterberg and James in chapter 2 above.
36  Cf. the section "Refuted and Ridiculed" in chapter 2 above.
37  Ludwig Wittgenstein, *Philosophical Remarks*, § 52.

language… This would hardly be understandable if it would not express experiences one can make in everyday life".[38]

So what are these experiences? I am coming to the conclusion of this chapter. Let me sum up my argument by saying that the metaphor of the flow of time is a specific, complex, figure of thought, synthesizing the experience of the passage of time as a physical force on the one hand, and the experience of the present as gradually receding into the past on the other. Both these experiences are veridical. The metaphor of the *flow of time* is a figure of thought expressing, in a unique way, an aspect of reality itself.

---

38  "[D]ie Rede vom Vergehen der Zeit [ist] keineswegs auf Philosophen beschränkt …, sondern [findet] in der gewöhnlichen Sprache in vielen Varianten Verwendung… Dies dürfte kaum zu verstehen sein, wenn darin nicht Erfahrungen Ausdruck verliehen wäre, die man im gewöhnlichen Leben machen kann." (Walter Mesch, "Wittgenstein über das Vergehen der Zeit", in Uwe Meixner and Peter Simons [eds.], *Metaphysik im postmetaphysischen Zeitalter: 22. Internationales Wittgenstein Symposium*, Kirchberg am Wechsel: ÖLWG, 1999, vol. 2, p. 47. See also my reference to Mesch in note 70 of chapter 4 above.)

# 6. Images in Conservative Education

Conservatism is a perennial human attitude and a constantly present cultural factor. As a consciously held theory it was however not formulated before the eighteenth century, and the expression "conservatism" itself was not in use before the 1830s. In the second section of the present chapter, under the heading "The Meaning of Conservatism", I will attempt both to convey a general idea of conservatism as well as to give a brief characterization of its three main historical phases: premodern, modern, and postmodern. Especially in its modern and postmodern phases, conservatism is tormented by paradoxes. My ultimate aim in the chapter will be to show that these paradoxes dissolve once the dominance of, and the exclusive focus on, verbal communication is supplanted by allotting a proper role to the pictorial – to mental and physical images, and to visual thinking. Setting the stage for my argument, in the first section below I offer some glimpses of the vastly rich literature, extending well back into the nineteenth century, on the visual mind – the visual as accompanying, or even serving as the basis of, the verbal, and as accompanied, or even based on, the motor. In the third section, drawing in particular on the ideas of the liberal-conservative thinker F. A. von Hayek, I will describe the main dimensions of what might be called a conservative concept of knowledge, characterizing knowledge as local, dispersed, and embedded in practice. The implications of such a concept of knowledge for the educational system under modern/postmodern conditions are spelled out in the fourth section. The fifth section, "Images and Conservatism", is divided into three subsections. In the first subsection, I strive to show that the pictorial as such tends to be conservative, basically because it provides a stable and rich representation of reality. The epistemological stance of conservatism is that of common-sense realism; common-sense realism assumes, correctly in my view, that images, in principle, convey what there really is. And it is by displaying what there really is that images, as I attempt to demonstrate in the second subsection, can successfully take over the role of verbally formulated traditions, spurious verbal formulas telling us what there once supposedly was, and telling us unconvincingly that that is how it should always be. In the third, last, subsection I explain why I believe that, from the point of view of postmodern conservatism, the image, namely the moving image, can fulfil a special role. The postmodern condition is one of fundamental uncertainty. Simulations bringing together vast amounts of data in an easily understandable animation are today our best instruments for dealing with a radically uncertain future.

## Visual Thinking

In the previous chapter of the present volume I already had occasion to quote a central passage by psychologist and art theorist Rudolf Arnheim, from his book *Visual Thinking*: "What makes language so valuable for thinking ... cannot be thinking in words. It must be the help that words lend to thinking while it operates in a more appropriate medium, such as visual imagery."[1] Recall that in the same book Arnheim relates images – mental images as well as a type of drawings expressing them – to *gestures*, pointing out that in gestures the visual is intrinsically bound up with motor, with "the kinesthetic experiences of pushing, pulling, advancing, obstructing".[2] Arnheim was a leading later-generation representative of the Gestalt school of psychology, adhering to the founding generation's view that one cannot experience images without experiencing the patterns of forces they embody and convey. He was, also, very much aware of the pioneering role of the German philosopher-psychologist Theodor Lipps here;[3] while on the broader topic of the visual mind he essentially drew on the work of Galton, Ribot, Binet, and Titchener.

Ribot, Galton, Binet, and in no small measure William James, were all impressed by the fact that thought processes obviously occur even in cases where they are paralleled neither by language, nor by conscious imagery.[4] It was the unconscious or half-conscious underlying motor dimension Hippolyte Taine alluded to when in 1870 he wrote: "beneath the incomplete image a dull agitation is going on, and as it were, a swarm of feeble impulses which usually sum themselves up in an expressive gesture, a metaphor, a visible summary".[5] And Binet, in the

---

1 Rudolf Arnheim, *Visual Thinking*, Berkeley: University of California Press, 1969, pp. 231 f.
2 *Ibid.*, p. 118, cf. note 13 in chapter 5 above.
3 As Arnheim wrote: "Lipps anticipated the Gestalt principle of isomorphism for the relationship between the physical forces in the observed object and the psychical dynamics in the observer" ("The Gestalt Theory of Expression", in Rudolf Arnheim, *Toward a Psychology of Art: Collected Essays*, Berkeley: University of California Press, 1966, p. 58).
4 Compare the brief discussion I gave in the section "The Visual and the Motor", in chapter 1 above.
5 Hippolyte Adolphe Taine, *De l'intelligence* (1870), here quoted from the English translation: *On Intelligence*, New York: Henry Holt, 1875, vol. I, p. 89.

concluding passage of his 1903 essay "Imageless Thought" must have referred, again, to the motor level when saying:

> I suppose that the word, like the sensory image, gives precision to the thought which, without these two aids – that of the word and that of the image – would remain very vague. – I even presume that it is the word and the image which contribute the most to making us conscious of our thoughts. Thought is an unconscious act of the mind which, to become fully conscious, necessitates words and images. No matter what difficulty we have in depicting a thought which is imageless – and it is only for this reason that I say thought is unconscious – it nevertheless exists. Thought constitutes, if one wishes to define it by its function, a directing organizing force which I would willingly compare (this is probably only a metaphor) to the vital force which, directing the physical-chemical properties, models the shape of beings and leads to their evolution...[6]

It is clearly impossible in the present brief sketch to give even a rudimentary overview of the intellectual history of the subject, but Figure 1 perhaps captures at least the most essential nodes and links.

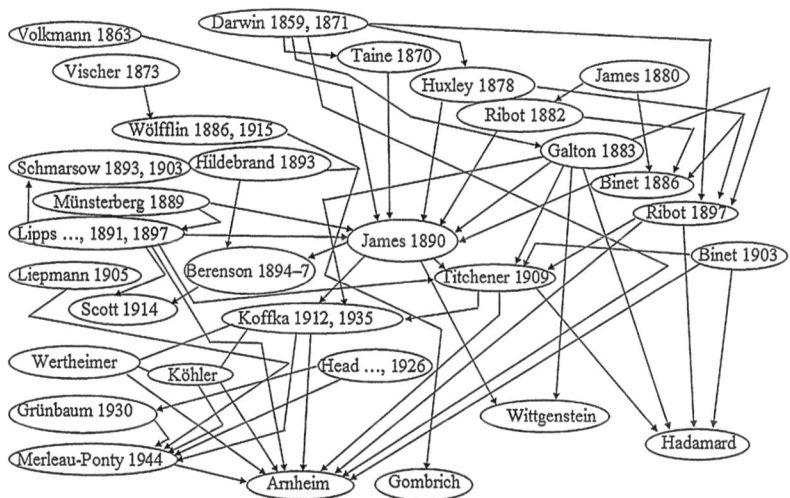

*Figure 1:* The visual and the motor. A network of influences in intellectual history

---

6   Alfred Binet, "La pensée sans images", here quoted from the English translation in *The Experimental Psychology of Alfred Binet*, ed. by Robert H. Pollack and Margaret J. Brenner, New York: Springer, 1969, p. 221.

Coming back to Arnheim and to German-language scholarship, let me here make just three more references. First, to Robert Vischer, who first elevated the term "Einfühlung" (subsequently elaborated by Lipps, and rendered as "empathy" by Titchener) into a technical term. "Stimuli in the thought domain", wrote Vischer, "can create sensitive as well as motor stimuli in the lower organs, and also the other way round. ... It is the whole body that is involved, the whole human body is seized", *der ganze Leibmensch wird ergriffen*.[7] Certainly the theory of the embodied mind is not a twentieth-century invention. Secondly, to a passage from Adolf Hildebrand's seminal 1893 book: "It is due to our vertical position with respect to the ground, and on the other hand to the horizontal position of our two eyes, that the vertical and the horizontal directions, as fundamental directions underlying all the others, are innate in us."[8] Thirdly, to the neurologist Grünbaum stressing, in 1930, that "*'pure' motoricity already possesses the capacity to elementary sense-giving ..., sense-giving as such goes back to motor connections*".[9] This might be, then, one of the contexts of intellectual history in which to see Arnheim when he makes, for instance, the observation: "the cross form as such can symbolize the conjunction of opposites, the action of centrifugal or centripetal forces, ... crossroads, the relation of vertical striving to horizontal stability, and so on."[10] Just like in the case of so many other age-old conventional symbols, stresses Arnheim, the cross as a perceptual pattern is the primary carrier of a broader meaning, while its conventional senses are narrower, and dependent on the former. A telling example, one Arnheim could not have been aware of at the time he wrote this remark but one that has become famous in the meantime, is the cross in Chauvet cave, a painting some 30 000 years old, discovered in 1994.[11] We recognize it as a symbol, and can give it an interpretation, without knowing what it precisely meant to the culture that created it.

---

7    Robert Vischer, *Über das optische Formgefühl: Ein Beitrag zur Ästhetik*, Leipzig, Hermann Credner, 1873, sect. 2.

8    Adolf von Hildebrand, *Das Problem der Form in der bildenden Kunst* (1893), 3rd., rev. ed., Strassburg: Heitz, 1901, p. 68.

9    A. A. Grünbaum, "Aphasie und Motorik", *Zeitschrift für die gesamte Neurologie und Psychiatrie*, vol. 130, nos. 1–3 (Berlin: Julius Springer, 1930), p. 394, italics in the original.

10   Rudolf Arnheim, *The Dynamics of Architectural Form* (1977), Berkeley: University of California Press, 2009, p. 209.

11   Jean-Marie Chauvet – Eliette Brunel Deschamps – Christian Hillaire, *Dawn of Art: The Chauvet Cave. The Oldest Known Paintings in the World* (1995). Epilogue by Jean Clottes. Foreword by Paul G. Bahn. Translated from the French by Paul G. Bahn. New York: Harry N. Abrams, 1996.

# The Meaning of Conservatism

Although he had a keen sense for the achievements and promises of modern art and architecture, Arnheim was no doubt a conservative. His conservatism was made up of two very different dimensions, an unusual and inspiring one, to which I will return shortly, and the customary backward-harking one, deploring contemporary "social conditions that atomize the human community into a mere aggregate of individuals or small groups", "the chaos of our present way of life", our "individualistic civilization".[12] It is this latter type of conservatism the Austrian novelist and essayist Robert Musil rejected when writing in 1923: "Having freed himself from all the old bonds, man is recommended to subject himself to them anew: faith, ... austerity, ... sense of national community, a concept of civic duty, and abandonment of capitalist individualism and all its attitudes. ... – The belief is that a decay has to be cured. – ... I can think of hardly any account which conceives of our present condition as a problem, a new sort of problem, and not as a solution that has miscarried."[13]

What Musil here points to is one of the fundamental paradoxes of conservatism as usually conceived. The demand that people should give up their present patterns of life, and return to the ways of some earlier age, is actually a revolutionary one, in need of argument or at least persuasion. If on the other hand conservatism is understood as the teaching that one should strive to preserve whatever norms and social conditions one happens to live under, we are again faced with a paradoxical doctrine indeed, one preaching different values according to different times and places. And yet another set of paradoxes emerges when conservatism

---

12 *The Dynamics of Architectural Form*, pp. 17 and 67. The passage on p. 17 begins with Arnheim complaining about "the visual, functional, and social chaos of modern life"; on p. 206 he refers, again, to "the prevailing individualism of our civilization". The term "civilization" to Arnheim's German ears clearly suggested something of the opposite of "culture", just as it did, say, to Thomas Mann, Oswald Spengler, or Ludwig Wittgenstein. In English of course the two terms are more often than not used as synonyms, cf. e.g. Franz Rauhut, "Die Herkunft der Worte und Begriffe 'Kultur', 'Civilisation' und 'Bildung'" (1951), *Germanisch-Romanische Monatsschrift* 34 (1953), pp. 81–91, and especially Wolfgang Schmidt-Hidding et al., *Kultur und Zivilisation* (*Europäische Schlüsselwörter*, vol. III), München: Max Hueber, 1967, see in particular pp. v–vi, 180 ff., 196 and 313 f.
13 Robert Musil, "Der deutsche Mensch als Symptom" (1923), in Robert Musil, *Gesammelte Werke*, ed. by Adolf Frisé, vol. 8, Reinbek bei Hamburg: Rowohlt, 1978, p. 1382, here quoted from the English translation in Kristóf [J. C.] Nyíri (ed.), *Austrian Philosophy: Studies and Texts*, München: Philosophia Verlag, 1981, p. 185.

is equated, as it almost invariably is, with "traditionalism". Traditions in the strict sense of the term are, as twentieth-century scholarship has exhaustively established, mechanisms for *preserving knowledge* – practices, techniques, as well as verbal knowledge – *in pre-literal cultures*.[14] It is simply misleading to speak of traditionalism where conditions of alphabetic literacy obtain. Hence it is misleading, too, to define conservatism, as Karl Mannheim does, as "primarily nothing more than traditionalism become conscious".[15] Mannheim chooses not to regard conservatism as "a phenomenon universal to all mankind".[16] When looking for a designation of the "general psychological attitude" ultimately underlying modern conservatism, he prefers Max Weber's term "traditionalism" to Lord Hugh Cecil's formula "natural conservatism".[17]

By contrast, as I indicated at the beginning of this chapter, one might well try to understand conservatism precisely as a perennial endeavour. I am coming back to Arnheim. In an essay written in 1969 he noted a contrast between, on the one hand, "British empiricist philosophy ... proudly asserting the dominion of the individual's views and judgments over the environment", and, on the other hand, the Gestalt view manifesting "respect for the structure of the physical world as it impinges upon the nervous system" and "affirming that it [is] man's task to find his own humble place in the world and to take the cues for his conduct and comprehension from the order of that world[,] ... demand[ing] of the citizen that he derive his rights and duties from the objectively ascertained functions and needs of society".[18] Humility, one's recognition of one's "humble place in the world" is, I take Arnheim to imply, a defining conservative stance. It is also, one should observe, a posture with a religious tinge.

Another point emerging from what Arnheim in this passage says is that one can in fact identify a constant task conservatism invariably faces. It is to comprehend the world as given, to acquire objective knowledge. Indeed it can be maintained that what conservatism in any historical age first and foremost strives to conserve is

---

14  For a survey, see my "Introduction: Notes towards a Theory of Traditions", in Kristóf [J. C.] Nyíri (ed.), *Tradition*, Wien: IFK, 1995, pp. 7–32 (accessible online at www.hunfi.hu/nyiri/Notes_towards_a_Theory_of_Traditions.pdf).
15  *From Karl Mannheim*, ed. by Kurt H. Wolff, New Brunswick: Transaction Publishers, 1993, p. 288. The quoted passage is from Mannheim's "Conservative Thought", an English translation based on his 1925 Heidelberg dissertation.
16  *From Karl Mannheim*, p. 280.
17  *Ibid.*, pp. 280 f.
18  Rudolf Arnheim, "Wertheimer and Gestalt Psychology" (1969), in Arnheim, *New Essays on the Psychology of Art*, Berkeley: University of California Press, 1986, p. 34.

actually *knowledge*, specifically the knowledge necessary to protect the life chances of future generations. However, such knowledge varies greatly, depending on the dominant information and communication technology of the age. In a preliterate culture, what society knows is limited to what people remember. Words, in a preliterate culture, are exclusively spoken or heard; knowledge has to be memorized through frequent repetition of texts the truth of which is taken to be indubitable due the fiction that they are passed down unchanged from generation to generation, with an ultimately divine origin. This, then, is the age of traditions, spanning the whole of premodernity, includeing also the centuries of manuscript culture, still dominantly oral.[19] Premodern conservatism struggles to safeguard the life of future generations by seeking to ensure the survival of the customs and beliefs of former generations. Modern conservatism by contrast, conservatism in the age of the printed press, cannot but recognize that change is inevitable. It attempts to slow it down, reduce its risks, by taking on the role of defending evolutionary social growth against the devastating influence of speculative theories. It emphasizes the knowledge embedded in the institutions and practices of society. This is Burke's line. Now postmodern conservatism, conservatism in the age of online networked communication, faces not only incessant inevitable change, but has to cope with shifts that are rapid and might be entirely unforeseeable. Postmodern conservatism, tormented by the paradox of preparing for what it cannot predict, has the task of continuously mobilizing, and keeping in readiness, the whole array of human knowledge. To be able to manage this, it has to have an adequate notion of what human knowledge really is like.

## The Conservative Concept of Knowledge

Burke's late-eighteenth-century views on knowledge as embedded in the institutions and practices of society were taken up and elaborated by Hayek in the 20th century. What Hayek has shown was that the knowledge necessary for society in order to maintain its economy, even in the case of a large-scale modern economy, emerges from, indeed is essentially upheld by, the practical experience society's individual members have with local conditions. It is knowledge distributed among individual market actors, mediated by the movement of prices, knowledge impossible to centralize. Now what holds for knowledge in the world of production, commerce, and services, appears to hold for knowledge generally, too. John Gray wrote of

---

19  For a more detailed discussion see my volume *Tradition and Individuality*, Dordrecht: Kluwer, 1992, esp. pp. 75 ff., compare also the prefatory passages on p. ix, *ibid.*

> Hayek's ... insight that all our theoretical, propositional or explicit knowledge presupposes a vast background of tacit, practical and inarticulate knowledge. Hayek's insight here parallels those of Oakeshott, Ryle, Heidegger, and Polanyi; like them he perceives that the kind of knowledge that can be embodied in theories is not only distinct from, but also at every point dependent upon, another sort of knowledge, embodied in habits and dispositions to act. Some of this practical knowledge is found in rules of action and perception imprinted in the nervous system and transmitted by genetic inheritance. But much of the significant part of the practical knowledge expressed in our dealings with each other is passed on mimetically, in the cultural transmission of traditions or practices...[20]

Let me note that when Gray uses the word "mimetic", he does not thereby allude to *visual* imitation. The issue of visuality, not to mention the idea of pictoriality, did not play a role in the history of conservative thought from Burke to Hayek. It is of course present in Burke's *Philosophical Inquiry into the Origin of Our Ideas of the Sublime and Beautiful*,[21] but even there visuality is deemed to be of secondary importance in comparison with the verbal. When Burke wrote that "poetry and rhetoric do not succeed in exact description so well as painting does; their business is, to affect rather by sympathy than imitation"[22], his point was not to highlight the power of pictorial representation, but to refute the position that words signify by depending on, or giving rise to, mental images.[23] Hayek, who as a young man had contemplated to become a psychologist rather than an economist, in 1952 published the book *The Sensory Order*, expressing views that came close to some of

---

20  John Gray, "Hayek as a Conservative", first published in *Salisbury Review* in 1983, reprinted in John Gray, *Post-liberalism: Studies in Political Thought*, London: Routledge, 1993, the quoted passage on p. 34. To Michael Polanyi's notion of "tacit knowledge" Gray repeatedly refers here. Our "explicit knowledge", he writes, "is only the visible surface of a vast fund of tacit knowing" (*ibid.*).
21  See e.g. the discussion "Why Visual Objects of Great Dimensions Are Sublime", *The Works of the Right Honourable Edmund Burke*, in twelve volumes, vol. 1, London: John C. Nimmo, 1887, pp. 217 f.
22  *Ibid.*, p. 257.
23  As he for instance puts it: "the most general effect" of words "does not arise from their forming pictures of the several things they would represent in the imagination" (*ibid.*, pp. 251 f.), and "in the ordinary course of conversation we are sufficiently understood without raising any images of the things concerning which we speak" (*ibid.*, p. 253). I have referred to Burke's theory of imageless thought in my talk "Tradition and Practical Knowledge" (1985), in Kristóf [J. C.] Nyíri and Barry Smith (eds.), *Practical Knowledge: Outines of a Theory of Traditions and Skills*, London: Croom Helm, 1988, pp. 26 f.

the tenets held by the Gestalt school.[24] However, he did share neither the school's focus on the visual,[25] nor its epistemological realism.[26] Also, he was apparently

---

24 Thus one of Hayek's starting points is: "We all readily recognize as the same tune two different series of tones, or as the same shape or figure structures of different size and colour. In all these instances groups of stimuli which individually may be altogether different do yet as groups evoke the same sensory quality or are classified by our senses as the same gestalt." (F. A. Hayek, *The Sensory Order: An Inquiry into the Foundations of Theoretical Psychology*, Chicago: The University of Chicago Press, 1952, p. 13.) Also, Hayek of course accepts the insight that "in perception we do not merely add together given sensory elements", and that "complex perceptions possess attributes which cannot be derived from the discernible attributes of the separate parts" – but cannot resist commenting that this "most general aspect of the problem of gestalt" had been discerned "even before the rise of the gestalt school", and "is by now recognized by practically all schools of psychology", *ibid.*, p. 76. Further, he arrives at the conclusion ("again in agreement with the views of the gestalt school", as he remarks in brackets) "that there is no substantial difference between the acts of 'sensation' and of 'perception'" (*ibid.*, p. 78). Finally, Hayek was strongly attracted, as was also the Gestalt school, to the motor approach to perception. As he puts it: "practically all sensory impulses are evaluated in the light of, or corrected for, simultaneous muscular activities"; there are "motor responses to sensory stimuli which ... might almost be described as part of the act of perception" – for example "the classical instance of the kinesthetic sensations connected with the focusing of the eye". Also, Hayek adds, "the proprioceptive reports of the body postures and movements designed to help perception" serve "as a sort of indispensable background for the proper evaluation of the stimulus" (*ibid.*, pp. 93 and 92).
25 Actually there are very few passages in the book which touch on visuality. Let me single out the one on p. 144, *ibid.*: "some people of the eidetic type appear to be able by recalling vivid images to discover details in them which they had not noticed at the time of the original experience". Hayek here adds the important remark: "But the memory images need not always to be more 'abstract' than current perceptions. ... there ... exists little justification for any sharp distinction between the 'concrete' picture supplied by sense perception and the 'abstractions' which are derived from the former by the higher mental processes (or between the complete picture of a unique situation built up by the 'senses' from fixed elements, and the abstract features which the 'intellect' singles out from the picture which is supposed to be given prior to any abstraction)."
26 The perception of Gestalt qualities does not, for Hayek, amount to a direct acquaintance with the structure of reality. His typical term is "approximation": a "gradual evolution of the mental order involves ... a gradual approximation to the order which in the external world exists between the stimuli evoking the impulses which 'represent' them in the central nervous system", *ibid.*, p. 107.

quite unaware of the function fulfilled by mental images and pictorial communication[27] in the constitution of knowledge as inherently bound up with practice.

Postmodern conservatism by contrast, committed to understanding the nature of knowledge in the digital networked age, clearly cannot avoid, and of course has access to the technological means, to come to terms with the issue of the pictorial. It has to come to terms, also, with the very issue of knowledge networks. In his recent book on conservatism, Kieron O'Hara notes that the World Wide Web is "a liberal idea – it is designed to allow information to flow easily. Its very structure makes it harder for authoritarian regimes to retain control of those areas of life that have migrated online." However, as O'Hara puts it, the web "has many conservative properties. ... it is not laid down by a central authority".[28] Now the point that the workings of the internet can indeed suggest the plausibility of a broadly conservative perspective on knowledge and society has been given a much stronger formulation by the Hungarian-born physicist Albert-László Barabási. The internet is made up of billions of nodes with just a few links, and a relatively small number of "hubs" with a great many links. It is through the hubs that smooth and swift communication is maintained between the rest of the nodes. In the harsh words of Barabási, there is a "*complete* absence of democracy" and of "egalitarian values" on the web.[29] The "vast majority of documents are hardly visible, since a highly popular minority has all the links". We do indeed have free speech on the web, writes Barabási. The chances are, however, that "our voices are too weak to be heard".[30]

## Conservatism and Education

Educating for a postmodern society from a conservative point of view, then, first of all demands raising a sophisticated awareness for the nature of online networks – a respect for their spontaneous growth, but also an ability to harness the possibilities they offer. Sustained success in coping with the net presupposes informal life-long learning. Informal learning is clearly a form of learning that accords with the fact that it is impossible to centralize knowledge.

---

27 In a telling passage Hayek speaks of "communication by language proper, as distinguished from communication by gestures, facial expression, etc." (*ibid.*, p. 135).
28 Kieron O'Hara, *Conservatism*, London: Reaktion Books, 2011, p. 268.
29 Albert-László Barabási, *Linked: The New Science of Networks*, Cambridge, MA: Perseus Publishing, 2002, p. 56.
30 *Ibid.*, p. 174.

Conservatives should encourage informal learning, but should maintain, or call for, decentralization in the domain of formal learning, too. As O'Hara puts it, "the conservative will be pleased to see the development of a strong autonomous school sector where decision-making about curricula and standards is devolved to the lowest possible level". A "good education system" should not be there "to fill perceived gaps in the workforce". Education, writes O'Hara, "needs to provide knowledge about the world" in the sense of offering "deep knowledge" of the *context*s of problems.[31] Such education cannot but be "challenging and testing", will not "at any cost" avoid putting pressure on children – but, points out O'Hara, "there is no evidence that children thrive educationally in environments that they themselves shape".[32]

An important instance readily coming to mind here is the issue of digital texts vs. hardcopy ones. Young people today will tend to move almost exclusively in the world of digital documents, ever less attracted to the printed book, and thumbing in notes, or punching away on the keyboard when it comes to longer texts, without taking care of printouts. Now while there are a great many wonderful new vistas opening up in the digital world, still, leaving hardcopy documents entirely behind seems in some respects like stepping back into a preliterate culture. The position I summed up some twenty years ago does not appear to have lost its validity:

> Just as speaking, as a rule, is less coherent than writing, a text composed on screen tends to be less coherent than a text composed in handwriting or on the typewriter. The reason for this is obvious. Maintaining coherence is a matter of comparing texts with each other, as well as of comparing one bit of a text with other bits of the same text. On screen such comparisons can be executed to a very limited extent only. Depending on the system used and the kind of display available, one, two, or even more documents can be viewed simultaneously; but of each document only a small segment will be exposed at a time.[33]

Conservative educationists should strive to preserve a level of hardcopy culture amidst the tide of a rising and promising digital culture. And – to come to the main and last point of this chapter – they should encourage exploiting the resources of the digital medium for the production and dissemination of visual images as the ultimate foundations of conservative practice and theory.

---

31 O'Hara, *op. cit.*, pp. 139 and 137.
32 *Ibid.*, p. 138.
33 Kristóf [J. C.] Nyíri, "Thinking with a Word Processor", in R. Casati (ed.), *Philosophy and the Cognitive Sciences*, Vienna: Hölder-Pichler-Tempsky, 1994, pp. 63–74, this passage on p. 70 (paper accessible online at www.hunfi.hu/nyiri/KRB93_TLK.htm).

# Images and Conservatism

## Images Conservative

Although images can be radically subversive, they have indeed been used, throughout history, as instruments for preserving the status quo. In his book *Augustus and the Power of Images* Paul Zanker provides a fascinating description of the way the penetration of Roman society by Greek art, from the 2nd century BC onward, played a part in dissolving traditional conditions; but he shows, also, how the new visual world that emerged at the time of Octavian's rule contributed to the permanent peace of the empire.[34]

Secondly, images are conservative in the sense that they preserve, in unchanging form, pictorial knowledge. To recall a very early instance: cave paintings served not only the purposes of ritual, religion, or art; they came into being as an answer to the felt need of storing and communicating knowledge. Discussing the tool-making revolution of the Upper Palaeolithic, John Pfeiffer refers to the enormous increase in complexity of the social world, to a veritable information explosion, which rendered inevitable the renewing of the "tribal encyclopedia".[35] And with the advent of the mechanical image – the photograph, the film – even some details became stored the recording of which had not been purposely intended. In fact, thirdly, as I suggested in the introductory passage of the present chapter, and again when citing Arnheim's "Wertheimer and Gestalt Psychology" essay,[36] the pictorial as such is conservative in the sense that it tends to show the world as given, the world as it really is. Images can be experienced, also, as expressing what might be called a higher reality – expressing meanings additional to, and beyond, their straightforward pictorial ones, meanings they point to, but do not display.[37] A famous example is Caspar David Friedrich's painting "The Wanderer above the Sea of Fog". The painting shows a lonely figure confronting nature in what appears to be deep reverence.

---

34  Paul Zanker, *Augustus und die Macht der Bilder*, München: Beck, 1987.
35  John E. Pfeiffer, *The Creative Explosion: An Inquiry into the Origins of Art and Religion*, Ithaca, NY: Cornell University Press, 1982, see esp. pp. 121 ff. and 185 ff. The expression "tribal encyclopedia" was coined by Eric Havelock; Pfeiffer's work, focussing on memory and the visual, in fact complements Havelock's theory of traditions (on Havelock see my "Introduction: Notes towards a Theory of Traditions", cf. note 14 above).
36  Cf. note 18 above.
37  Compare my paper "Images in Natural Theology", in Russell Re Manning (ed.), *The Oxford Handbook of Natural Theology*, Oxford: Oxford University Press, 2013, esp. pp. 586 ff.

Giving expression to reverence by indirect visual means is a topic Arnheim repeatedly returns to in his *The Dynamics of Architectural Form*. A notable passage: "the very nature of religion and its tasks are now so open to question that their external expression is no longer governed by reliable standards. ... all the more rewarding [are] those examples of church architecture that succeed in translating dignity and spiritual devotion into twentieth-century idioms". Even the late-modern architect, suggests Arnheim, might achieve a "reinforcing [of] deep-seated spiritual connotations".[38] A piece of architecture Arnheim apparently regarded as a gratifying example is Le Corbusier's Chapel of Notre Dame du Haut.[39] And a work he definitely singled out is Mario Botta's church in Mogno, Switzerland. "In religious architecture", Arnheim wrote, "a good designer such as Mario Botta gave up most of the literal applications of tradition, not to ignore them but to probe once again the deeper core of human feeling and thought".[40]

## From Traditions to Images

Some thirty or forty years ago I have put together a theory of traditions which I thought was based on the philosophy of the later Wittgenstein.[41] With hindsight, I today realize that it was based, rather, on a one-sided interpretation of that philosophy, an interpretation doubtlessly made possible by the state of Wittgenstein editions as we had them at that time, presenting Wittgenstein straightforwardly as a linguistic philosopher.[42] The argument I was most comfortable with when

---

38   Arnheim, *The Dynamics of Architectural Form*, pp. 206 and 210.
39   Cf. *ibid.*, pp. 106 f.
40   See Rudolf Arnheim, "Notes on Religious Architecture" (1993), in Rudolf Arnheim, *The Split and the Structure: Twenty-eight Essays*, Berkeley, CA: The University of California Press, 1996, p. 61. I am indebted to Arnheim expert Ian Verstegen for drawing my attention to this essay, and for a number of insightful comments.
41   Suggesting, way back in 1976, not only that "Wittgenstein's so-called later philosophy is the embodiment of a conservative-traditionalist view of history", but also that "this philosophy in fact provides a logical foundation for such a view" (Kristóf [J. C.] Nyíri, "Wittgenstein's New Traditionalism", *Acta Philosophica Fennica*, vol. 28, nos. 1–3, pp. 503–512, this passage on p. 503).
42   Referring to his *Philosophical Investigations* as we then knew it, I felt it was possible to ascribe to Wittgenstein the view: "language-games, i.e. forms of life, have to be accepted, ... they are what is *given*... In any endeavour to criticize a given linguistic tradition, only another linguistic tradition can serve as a standard" ("Wittgenstein's New Traditionalism", p. 509).

advancing Wittgenstein as a persuasive traditionalist pertained to the domain of elementary mathematics. Believing to speak for Wittgenstein, I wrote:

> Two and two are four, and the only explanation we can give here is that *this is the way we count*. Arithmetical knowledge is based on a conformity in behaviour that is not replaceable by any kind of insight. More generally, traditionalism as here conceived maintains that in the absence of indubitable truths of fact and value there can be no communication, argumentation, or discussion, that society is *held together* by the uniform acceptance of such truths; and that it is education in the family and in the school that has to confer the proper authority upon these truths.[43]

A similar formulation that I attempted:

> The concept of … the human subject acting in accordance with the light of his reason, sovereign within his own mental world, reveals itself as absurd in the face of the realization that the meaning of a word is not a mental image, but the use to which the word is put; thinking, believing, expecting, hoping, and so on, are not private mental processes; mathematical insight is grounded in exercise, in drill… [As] Wittgenstein wrote: "Counting (and that means: counting like *this*) is a technique that is employed daily in the most various operations of our lives. And that is why we learn to count as we do: with endless practice, with merciless exactitude; that is why it is inexorably insisted that we shall all say 'two' after 'one', 'three' after 'two', and so on" [Wittgenstein, *Remarks on the Foundations of Mathematics*, Part I, § 4]. This conception of mathematical insight and of the ways in which arithmetic is learned, is rooted in the same psychological attitude as Wittgenstein's general conception of education. The latter may be illustrated, for example, by his remark: "When you say NO to a child, you should be like a wall and not like a door".[44]

Clearly, Wittgenstein did see a connection between rote learning and the acquisition of the ability to count. The error I have made was not to realize that he saw a very different kind of connection, too: the one holding between arithmetical truths and their visualizations. The error, indeed the blunder, easy to make at the time, was not to open my eyes to Wittgenstein's philosophy of images. Wittgenstein's manuscripts contain innumerable drawings and diagrams, most of them in fact illustrating points he made in the domain of the philosophy of mathematics, but

---

43 Kristóf Nyíri, "Szabadpiac és tekintélyelvű társadalom: Angolszász liberális-konzervatív elméletek" [The free market in an authoritarian society: Anglo-Saxon liberal-conservative theories], *Világosság*, August–September 1981, pp. 534–540, the translated passage on p. 540.

44 Quoted from the chapter "Wittgenstein 1929–31: Conservatism and Jewishness", in my volume *Tradition and Individuality* (cf. note 19 above), pp. 15 and 117. The chapter was based on a paper I originally published in German in 1982.

only a fraction of them published in the printed volumes edited by his literary executors. A particularly interesting example (Figure 2):

„... dies als Beweis von 3 + 2 = 5

$$\cdot \cdot \cdot | \quad \cdot \cdot \cdot | \quad \cdot \cdot \cdot |$$

(Oder kinematographisch vorgeführt.)"

*Figure 2:   Visual proof of 3 + 2 = 5. From Wittgenstein's MS 118, p. 65r (1937)*

Wittgenstein here suggests a way to prove the equation 3 + 2 = 5. The proof would consist in drawing a series of pictures, or in the "cinematographic" presentation of the same series – an *animation*. No wonder this idea did not make it into the printed editions. In mathematics, the first half of the twentieth century was still very much characterized by the visualization Angst that had emerged in the nineteenth.[45] Today this fear is receding. Here, again, Arnheim was well ahead of his time. It is not tradition but perceptual grasp that can best teach us the fundamentals of mathematics; and it is not tradition but perception – most importantly visual cognition – that tells us what reality is like.

## Images and the Unknown Future

The task of postmodern conservatism is to create conditions in which the knowledge necessary to maintain the life of future generations is optimally preserved. However, the postmodern conservative is painfully aware of the fact that the future cannot be predicted. Now both our inner mental imagery and the visible world surrounding us consist of *moving* images – still images being extreme cases of moving ones. The moving image preserves and shows, tells, narrates, but also foreshadows. In an animation the unknown future can, experimentally, be brought to life; an animation built on millions of data can well prove to be a successful simulation. It is the image that solves the paradox of modern conservatism, and it is the moving image that appears to be the most effective cognitive device to alleviate the paradox of postmodern conservatism.

---

45   Cf. the section "Visualization in Mathematics" in chapter 1 above.

# 7. Time and Image in the Theory of Gestures

As I indicated in the first chapter of the present volume, in the subsection "The Visual and the Motor", as well as in the section "Visual Thinking" in the previous chapter above, towards the end of the nineteenth century there emerged a psychological position according to which it is the whole body, the entire motor system, including facial expressions and bodily gestures, that underlies not just emotions, but also abstract thought. Meaning, both emotional and cognitive, should be conceived of as primordially grounded, and ultimately embodied, in the motor dimension. This psychological perspective was definitely conducive to inspiring the late-nineteenth-century and early-twentieth-century interest in the language of gestures, an interest that is today once more vivid.

One can speak about gestures, and about languages of gestures, in at least four, partly of course overlapping, senses. First, as referring to the natural language of deaf-mutes, today forming the basis of a great number of officially recognized sign languages, such as ASL (American Sign Language), or DGS (Deutsche Gebärdensprache). Secondly, in the sense of the hypothesis – an hypothesis to which observations on the language of deaf-mutes, too, might lead – that the original language of humankind was a language of gestures preceding vocal language. Thirdly, the past few decades have witnessed the emergence of increasingly extended research on the interplay of talk and spontaneous gesture. And fourthly, we are acquainted with various cultures of handed-down, conventional gestures, such as that of the Neapolitans, or of North American Indians, or say of the language of gestures of the Cistercians.

My first attempt to come to terms with the issue of gestures was in a paper I wrote in 2002.[1] I there relied in particular on a formulation by the neurologist Macdonald Critchley, going back to 1939, according to which there is a "'natural sign-language' of the deaf and dumb [which is] is largely unfamiliar to outsiders and indeed many are unaware of its very existence. ... Even very young deaf-mutes

---

1   Kristóf Nyíri, "Pictorial Meaning and Mobile Communication", in Kristóf Nyíri, ed., *Mobile Communication: Essays on Cognition and Community*, Vienna: Passagen Verlag, 2003, pp. 157–184.

communicate freely with each other and the presence of this natural sign-language at an age prior to their receiving systematic instruction points to an 'instinctive' or at least a primitive type of symbolization."[2] I took over from Critchley some photos, too, illustrating universal gestures of deaf-mutes on the one hand (Figure 1), and culturally specific, conventional gestures on the other (Figure 2). Also, I referred at some length to William Stokoe, who at the time was perhaps the best-known representative of the position arguing for a priority of the language of gestures. In his last book, *Language in Hand*, published in 2001, Stokoe summarized his earlier arguments. One of his fascinating theses was that not only the *semantics* of verbal languages (the word meanings they carry), but also their *syntax*, in particular the subject–predicate structure, is prefigured in gestures. Handshapes (motionless, or with small, repeated motions) function as names, *moving* handshapes function as verbs. Together, they amount to *sentences*.[3]

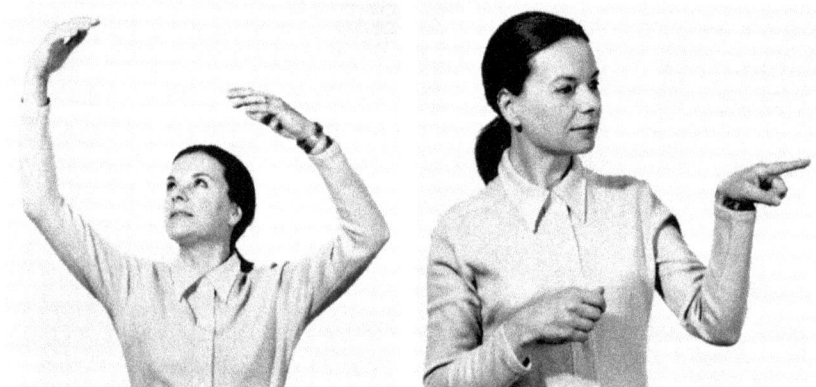

*Figure 1:*     *The natural gesture language of the deaf and dumb. Sign on the left indicates "heaven", on the right "over there". (After Critchley)*

---

2    Macdonald Critchley, "Kinesics; Gestural and Mimic Language: An Aspect of Non-Verbal Communication" (a paper based in part on Critchley's 1939 book *The Language of Gesture*, London: Arnold, 1939), in his collection *Aphasiology and Other Aspects of Language*, London: Edward Arnold, 1970, pp. 305 f. – Among the earlier studies Critchley specifically refers to in his "Kinesics..." paper is David Efron, *Gesture and Environment*, New York: King's Crown, 1941. Efron's book has in the meantime become one of the classics of the topic (new ed. 1972: *Gesture, Race and Culture*, The Hague: Mouton).

3    William C. Stokoe, *Language in Hand: Why Sign Came Before Speech*, Washington, D.C.: Gallaudet University Press, 2001, pp. xiii and 12 f.

*Figure 2:  Italian gestures: Approval–Contentment–Excellent!–I insist. (After Critchley)*

## The Theory of Gestures: A Nutshell History

Now a minimally complete history of the theory of gestures – a history of which I will, here too, provide an only very rudimentary sketch – should clearly begin with Plato's *Cratylus*, referring to the lines: "Suppose that we had no voice or tongue, and wanted to communicate with one another, should we not, like the deaf and dumb, make signs with the hands and head and the rest of the body? ... We should imitate the nature of the thing; the elevation of our hands to heaven would mean lightness and upwardness; heaviness and downwardness would be expressed by letting them drop to the ground."[4] Next I assume I would have to quote Quintilian as saying: "though the peoples and nations of the earth speak a multitude of tongues, they share in common the universal language of the hands"[5] – then taking a leap to the 17th century, making a detour round George Dalgarno,[6] but pausing briefly to recall the understandable interest Leibniz had in the language of gestures as a possible universal sign language.[7] By contrast, a more detailed narrative should be

---

4   *Cratylus*, 422e–423a, transl. by Benjamin Jowett.
5   Quintilian, *Institutio oratoria*, XI, 3, 87, transl. H. E. Butler.
6   Author of *Didascalocophus, or the Deaf and Dumb Man's Tutor*, Oxford: 1680.
7   Cf. e.g. Garrick Mallery, "Sign Language among North American Indians Compared with that among Other Peoples and Deaf-Mutes", First Annual Report of the Bureau of Ethnology to the Secretary of the Smithsonian Institution, 1879–1880, Washington: Government Printing Office, 1881, pp. 288, 349 f. and 360; Karl Sittl, *Die Gebärden der Griecher und Römer*, Leipzig: Teubner, 1890, p. 5; Wilhelm Wundt, *Völkerpsychologie: Eine Untersuchung der Entwicklungsgesetze von Sprache, Mythus und Sitte*, vol. I: *Die Sprache*, 2., rev. ed., Leipzig: Engelmann, 1904, p. 151. The chapter of Wundt's work discussing gestures has been published in an English translation: Wilhelm Wundt, *The Language of Gestures*, The Hague: Mouton, 1973, the reference to Leibniz here find on p. 70.

allotted to the 18th century. Not perhaps because of Vico, whose *Scienza nuove*, first published in 1725,[8] for a long time "went virtually unnoticed outside of Naples"[9], due not least to what has been called "the obscurity of his message"[10] – the message, in the case of our present topic, amounting to just two passages (repeated twice with slight variations) in the course of the entire book: "Mutes make themselves understood by gestures or objects that have natural relations with the ideas they wish to signify", and: "Since it has been demonstrated that the first gentile nations were all mute in their beginnings, they must have expressed themselves by gestures or by physical objects having natural relations with their ideas".[11] Nor has Rousseau contributed that much to the theory of gestures. Corballis is of course right when he finds the passage "Words would seem to have been necessary to establish the use of words"[12] an important formulation of the paradox bedevilling any theory that wants to explain the emergence of language without having recourse to the significance of gestures. But the conclusion Rousseau draws from this paradox in his *Origin of Languages*, namely that "Although the language of gesture and spoken language are equally natural, still the first is easier and depends less upon conventions",[13] is a rather pale one, and at any rate the essay was never published by him.

---

8    The third edition – the final one in Vico's lifetime – being published in 1744. This is the edition that served as the basis of the first ever English translation: Giambattista Vico, *The New Science*, Ithaca, N.Y.: Cornell University Press, 1948.

9    Marcel Danesi, *Vico, Metaphor, and the Origin of Language*, Bloomington: Indiana University Press, 1993, p. viii.

10    Bertrand Russell, *The Wisdom of the West: A Historical Survey of Western Philosophy in Its Social and Political Setting*, Garden City, N.Y.: Doubleday, 1959, p. 207. It should be pointed out however that the text of *Wisdom of the West* was actually drafted by the editor Paul Foulkes, on the basis of Russell's *A History of Western Philosophy*. Russell read it in proof, cf. Carl Spadoni, "Who Wrote Bertrand Russell's *Wisdom of the West?*", *Papers of the Bibliographical Society of America*, vol. 80, no. 3 (1986). The *History of Western Philosophy* makes no reference to Vico.

11    *The New Science*, §§ 225 and 434 (Engl. transl. pp. 68 and 127), see also §§ 401 and 431 (Engl. transl. pp. 114 and 125).

12    This is the translation Corballis himself gives of the wording "la parole paraît avoir été fort nécessaire, pour établir l'usage de la parole", in Rousseau's *Discours sur l'origine et les fondements de l'inégalité parmi les hommes* (1754), see Michael C. Corballis, *From Hand to Mouth: The Origins of Language*, Princeton, NJ: Princeton University Press, 2002, p. 42. The translation by G. D. H. Cole, as also the recent one by Johnston, seems to miss the essential point.

13    Jean-Jacques Rousseau, *Essay on the Origin of Languages*, transl. by John H. Moran, New York: F. Ungar, 1966, p. 6.

It was the philosopher Condillac and the educationalist de l'Épée whose work made the 18th century into a turning point in the history of the theory of gestures. Condillac's *Essai sur l'origine des connaissances humaines*, published in 1746, with a first English translation (*An Essay on the Origin of Human Knowledge*) appearing in 1756, formulates a detailed hypothesis on how a language of gestures could have preceded vocal language.[14] The Abbé de l'Épée from the 1750s onward became the founder of a unique teaching method for deaf children, based on their own common-spontaneous gestural language, "a natural sign language", as l'Épée saw it in his 1776 book *L'institution des sourds et muets*.[15] In the book l'Épée referred specifically to gestures signalling the passage of time – the past, the present, and the future. For instance, he found that "the pupils he encountered signified that an action or event was past by throwing the hand back beside the shoulder once or repeatedly".[16] A similar gesture with a similar meaning one encounters today say in DGS, the recognized German sign language. I will come back to this topic in the final section of the present chapter.

L'Épée and his school – one should here name, in particular, his immediate successor, the Abbé Sicard – soon gained wide influence both in Europe (most importantly perhaps in Germany) and in North America.[17] Still, in the 19th century, which I have now arrived at with my rudimentary narrative, the position that the language of gestures historically precedes vocal language, and that the former might take on a new pedagogical role, was far from having become a majority one. To be sure, in 1832 there appeared, and soon became rather widely known, the work *Gesture in Naples and Gesture in Classical Antiquity* by Andrea de Jorio, in which the author argues, if not for the priority, but at least for the unique expressive value, and a continuity throughout the centuries, of the south Italian gesture

---

14   Corballis provides an appreciative description of Condillac's main argument in his *From Hand to Mouth*, pp. 64, 102 f. and 126 f.
15   Charles Michel de l'Épée, *L'institution des sourds et muets, par la voie des signes méthodiques*, Paris: Nyon l'aîné, 1776, p. 126: "la langue naturelle des signes", see also the expression "Signes naturels" on the title page of the book.
16   William C. Stokoe, "Sign Language Structure: An Outline of the Visual Communication Systems of the American Deaf" (1960), reprinted in the *Journal of Deaf Studies and Deaf Education*, vol. 10, no. 1 (2005), pp. 3–37, the quoted passage on p. 5.
17   Roch-Ambroise Sicard wrote the important book *Cours d'Instruction d'un sourd-muet de naissance* (Paris: Le Clere, 1803). On l'Épée, Sicard, and their impact in America see the classic 1960 paper by Stokoe, referred to in the previous note.

language.[18] In 1838 there was published the wide-ranging and deep study *Ueber die Taubstummen und ihre Bildung* by Eduard Schmalz,[19] in 1853 the book *Ueber Taubstumme, Taubstummen-Bildung und Taubstummen-Anstalten* by Otto Friedrich Kruse,[20] and in 1865 Tylor's seminal work *Researches into the Early History of Mankind*, referring to Sicard as well as both to Schmalz and Kruse, and in great detail to "the Berlin Deaf-and-Dumb Institution"[21], discussing in the first three chapters "the gesture-language", and in the next two the topics of "gesture-language and word-language" and "picture-writing and word-writing".

Let me here quote at some length from Tylor. This is how he introduces the issue:

> The mother-tongue (so to speak) of the deaf-and-dumb is the language of signs. The evidence of the best observers tends to prove that they are capable of developing the gesture-language out of their own minds without the aid of speaking men. Indeed, the deaf-mutes in general surpass the rest of the world in their power of using and understanding signs, and for this simple reason, that though the gesture-language is the common property of all mankind, it is seldom cultivated and developed to so high a degree by those who have the use of speech, as by those who cannot speak, and must therefore have recourse to other means of communication.[22]

Tylor then cites Schmalz as pointing out that there are "many signs which we indeed do not use in ordinary life, but which the deaf-and-dumb child uses, having no means of communicating with others but by signs. These signs consist principally in drawing in the air the shape of objects to be suggested to the mind, indicating their character, imitating the movement of the body in an action to be described, or the use of a thing, its origin, or any other of its notable peculiarities."[23] Tylor entirely endorses the view that the basis of deaf-mute communication is pantomimic. Also, he assumes, even if the formulation he uses is a restrained one, that there is no thinking without communication, "without some means of outward expression" –

---

18  *La mimica degli antichi investigata nel gestire napoletano*. The English translation, with an excellent introduction by the translator Adam Kendon, has been recently published: Andrea de Jorio, *Gesture in Naples and Gesture in Classical Antiquity*, Bloomington: Indiana University Press, 2002.
19  Eduard Schmalz, *Ueber die Taubstummen und ihre Bildung*, Dresden und Leipzig: Arnoldische Buchhandlung, 1838.
20  Otto Friedrich Kruse, *Über Taubstumme, Taubstummen-Bildung und Taubstummen-Anstalten: Nebst Notizen aus meinem Reisetagebuche*, Schleswig: Bruhn, 1853.
21  Edward B. Tylor, *Researches into the Early History of Mankind and the Development of Civilization* (1865), Boston: Estes & Lauriat, 1878, p. 20.
22  *Ibid.*, pp. 17 f.
23  *Ibid.*, p. 18. Tylor is here translating a passage from p. 267 of the book by Schmalz.

while of course the deaf-mute can very well think without speech in the sense of "articulate sounds".[24] Tylor's unequivocal, radical, even if not explicitly stated conclusion: we clearly encounter thinking built up solely by movements and images of movements. A second obvious conclusion however, that of the historical priority of the language of gestures, is one Tylor clearly abstains from. "The idea that the Gesture-Language represents a distinct separate stage of human utterance, through which man passed before he came to speak, has no support from facts", he writes.[25]

The fundamental argument for this obvious conclusion – the argument foreshadowed by Rousseau's paradox quoted above with a reference to Corballis – was memorably formulated by the American political figure Amos Kendall in his speech at the inauguration of the College for the Deaf and Dumb in Washington DC, in 1864. "We read", said Kendall, "that Adam named the beasts and birds. But how could he give them names without first pointing them out by other means? How could a particular name be fixed upon a particular animal among so many species without some sign indicating to what animal it should thereafter be applied?"[26] In the course of human phylogeny, Kendall indicated, it was the language of gestures, and not verbal language, which introduced conceptual order into the episodic imagery of pre-linguistic thought. The reference to Adam, five years after the publication of Darwin's *The Origin of Species*, I rather take to be an ironical one.

## Darwin on the Expression of Emotions

Darwin himself markedly contributed to the theory of bodily and facial gestures with his 1872 book *The Expression of the Emotions in Man and Animals*. The book's main proposition: gestures have an evolutionary basis, they originate in concrete bodily reactions to events in the surrounding environment, to danger, threat, and so on. Let me here focus on gestures of affirmation and negation. In an introductory passage of his book, in the first chapter, Darwin cites with approval the observation that "[a] man ... who vehemently rejects a proposition, will almost certainly shut his eyes or turn away his face; but if he accepts the proposition, he will nod his

---

24 *Ibid.*, p. 14.
25 *Ibid.*, p. 15. Tylor returned to the topic of gesture-languages in his book *Anthropology: An Introduction to the Study of Man and Civilization*, London: Macmillan and Co., 1881.
26 I am quoting after David F. Armstrong – Sherman E. Wilcox, *The Gestural Origin of Language*, New York: Oxford University Press, 2007, p. 8.

head in affirmation and open his eyes widely. The man acts in this latter case as if he clearly saw the thing, and in the former case as if he did not or would not see it."

In the chapter dealing with disdain, contempt, disgust, and affirmation and negation, Darwin quotes Tylor's *Researches into the Early History of Mankind* to explain how the gesture "snapping one's fingers", indicating contempt, becomes intelligible once "we notice that the same sign made quite gently, as if rolling some tiny object away between the finger and thumb, or the sign of flipping it away with the thumbnail and forefinger, are usual and well-understood deaf-and-dumb gestures, denoting anything tiny, insignificant, contemptible". It seems, Tylor concludes, "as though we had exaggerated and conventionalized a perfectly natural action, so as to lose sight of its original meaning". Some passages later Darwin offers an interim summary. "We have now seen that scorn, disdain, contempt, and disgust are expressed in many different ways, by movements of the features, and by various gestures; and that these are the same throughout the world. They all consist of actions representing the rejection or exclusion of some real object which we dislike or abhor...". A few pages further there follows the section "*Signs of affirmation or approval, and of negation or disapproval: nodding and shaking the head.*" He was "curious to ascertain", Darwin here writes,

> how far the common signs used by us in affirmation and negation were general throughout the world. These signs are indeed to a certain extent expressive of our feelings, as we give a vertical nod of approval with a smile to our children, when we approve of their conduct; and shake our heads laterally with a frown, when we disapprove. With infants, the first act of denial consists in refusing food; and I repeatedly noticed with my own infants, that they did so by withdrawing their heads laterally from the breast, or from anything offered them in a spoon. In accepting food and taking it into their mouths, they incline their heads forwards. ... It deserves notice that in accepting or taking food, there is only a single movement forward, and a single nod implies an affirmation. On the other hand, in refusing food, especially if it be pressed on them, children frequently move their heads several times from side to side, as we do in shaking our heads in negation. Moreover, in the case of refusal, the head is not rarely thrown backwards, or the mouth is closed, so that these movements might likewise come to serve as signs of negation.

Three remarks. First, that Darwin's explanatory pattern, the tracing back of an emotion to the actual behaviour on which it is based, unmistakably anticipates the James–Lange theory of emotions. As the classic summary formula given by James runs: "*the bodily changes follow directly the perception of the existing fact, and ... our feeling of the same changes* IS *the emotion*".[27] Secondly, that I am

---

27 William James, *The Principles of Psychology* (1890), London: Macmillan & Co., 1901, vol. II, p. 449. A source of inspiration for James here is Carl Georg Lange, *Über*

here mainly concerned with preparing the ground for what I will attempt to claim when I come to the topic gestures of time in the last section of the present chapter: natural gestures allow us to infer that what they embody is the experiencing of something real. Thirdly, that obviously there are numerous different patterns of behaviour from which gestures of affirmation and negation can emerge, patterns linked to each other by family resemblances. Garrick Mallery, in his fundamental, very extensive study "Sign Language among North American Indians Compared with that among Other Peoples and Deaf-Mutes", published in 1881,[28] provides a wide variety of illustrations; similarly Karl Sittl, in his 1890 book *Die Gebärden der Griecher und Römer*.[29] Nor are the corresponding signs in today's gesture languages of the deaf restricted to a mere nodding or shaking of the head.

## From Wundt to Corballis

I have now, with this rudimentary history of the theory of gestures, at long last arrived at the 20th and 21st centuries. Volume I of Wilhelm Wundt's *Völkerpsychologie*, published in 1900, contains an absolutely brilliant discussion of the subject. For Wundt, gesture language has "an originality and naturalness such as speech neither possesses today nor has ever had in any forms hitherto uncovered by linguistics"; he highlights the merits of the view according to which "gestural communication is the original means of communication. This would mean that gesture, as the natural aid of communication, preceded spoken language";[30] but points out, too, that "systems of signs that have arisen in spatially separate environments and under doubtlessly independent circumstances are, for the most part, very similar or indeed closely related; this, then, enables communication without great difficulty between persons making use of gestures. Such is the much-lauded universality of gestural communication."[31] Wundt can conceive of a mental makeup where "all powers of consciousness are concentrated on thought in terms of gestural images only".[32] And it is not only concrete, but also symbolic gestures that "will reach back in the earliest, if not the beginning stages of the system. The over-all character of the

---

*Gemüthsbewegungen: Eine Psycho-physiologische Studie*, Leipzig: Verlag Theodor Thomas, 1887.
28 Cf. note 7 above.
29 Cf. note 7 above, see esp. p. 82 in Sittl's book.
30 Wundt, *The Language of Gestures* (cf. note 7 above), p. 56.
31 *Ibid.*, pp. 58 f.
32 *Ibid.*, p. 60.

symbolic gesture ... consists of transmitting the concept to be communicated from one field of perception to another"[33]. The basic idea of today's conceptual metaphor theory, including this theory's attention to visual metaphors, is clearly there in Wundt's work. The issue of gesture languages was very much present in Ogden and Richards' classic 1923 volume, *The Meaning of Meaning*. "Words, whenever they cannot directly ally themselves with and support themselves upon gestures", they wrote, "are at present a very imperfect means of communication."[34] A magnificent attempt at a synthesis of the theories of meaning, motoricity and gestures is Merleau-Ponty's *Phénoménologie de la perception*, published in 1945. Let me just quote two passages from this work. The first, on emotion and gesture: "Faced with an angry or threatening gesture, I have no need, in order to understand it, to recall the feelings which I myself experienced when I used these gestures on my own account. ... I do not see anger or a threatening attitude as a psychic fact hidden behind the gesture, I read anger in it. The gesture *does not make me think* of anger, it is anger itself."[35] The second, a version of the argument we have already encountered in the formulations of Rousseau and Kendall:

> was not the communication of the elements of language between the "first man to speak" and the second necessarily of an entirely different kind from communication through gesture? This is what is commonly expressed by saying that gesture or emotional pantomime are "natural signs", and the word a "conventional sign". But conventions are a late form of relationship between men; they presuppose an earlier means of communication, and language must be put back into this current of intercourse.[36]

---

33  *Ibid.*, p. 74, and let me here quote the second part of the passage in the original German, too: "Der allgemeine Charakter der symbolischen Gebärde besteht ... darin, daß sie die auszudrückenden Vorstellungen aus einem Anschauungsgebiet in ein anderes überträgt". – Wundt's work on gestures was extensively discussed by George Herbert Mead, see his *Mind, Self and Society*, Chicago: The University of Chicago Press, 1934 (a posthumous volume based on lecture notes; Mead himself published two papers on Wundt early in the century).

34  C. K. Ogden – I. A. Richards, *The Meaning of Meaning: A Study of the Influence of Language upon Thought and of the Science of Symbolism*, London: Routledge & Kegan Paul, 1923, ch. I.

35  Maurice Merleau-Ponty, *Phenomenology of Perception*, London: Routledge & Kegan Paul, 1962, p. 184.

36  *Ibid.*, p. 187. The English edition has "natural convention" instead of "conventional sign" ("signe conventionnel") – clearly a slip of the typewriter.

A new interest in the language of gestures emerges in the humanities from the 1960s onward. The literature is vast, and I can certainly not attempt to give a survey of it here.[37] Outstanding is the book *From Hand to Mouth: The Origins of Language* by Michael Corballis, published in 2002. Corballis unambiguously sides with the thesis that "human language evolved first as a system of manual gestures", with "communicative gestures emerg[ing] from actions on the physical world and ... then adapted and conventionalized".[38] Referring to Merlin Donald's notion of a "mimetic stage" in human evolution,[39] Corballis writes: "The actions involved in making or using tools could have come to represent the tools themselves, or perhaps the hands and arms were used to depict the actual shapes of things." Gestures were primordially iconic, but tended to condense into symbols. Today, too, "[s]tudies of deaf children inventing their own homesign ... suggest that signs are initially coined for their resemblances to what they represent but are later adapted to a more arbitrary form. ... it is the early gestures", runs the answer

---

37  But let me at least mention some of the most notable items, before coming (or coming back to, cf. notes 12 and 14 above) to Corballis. To Stokoe's 1960 paper and 2001 book I have referred to in notes 16 and 3 above, to the Armstrong–Wilcox book in note 26, an essential item belonging to this cluster is David F. Armstrong – William C. Stokoe – Sherman E. Wilcox, *Gesture and the Nature of Language*, Cambridge: Cambridge University Press, 1995. Adam Kendon, author of the introduction to the English translation of de Jorio's book (cf. note 18 above), has published the two seminal essays, "Some Relationships between Body Motion and Speech" (in A. Siegman and B. Pope, eds., *Studies in Dyadic Communication*, New York: Pergamon, 1972, pp. 177–210) and "Gesticulation and Speech: Two Aspects of the Process of Utterance" (in Mary Ritchie Key, ed., *The Relationship of Verbal and Nonverbal Communication*, The Hague: Mouton, 1980). A crucially important paper, singled out by Corballis, too, is Gordon G. Hewes, "Primate Communication and the Gestural Origin of Language", *Current Anthropology*, vol. 14, no. 1–2 (February–April 1973), pp. 5–24. Two influential books by David McNeill, on the interdependence of vocal language and spontaneous gesturing, are *Hand and Mind: What Gestures Reveal about Thought* (Chicago: The University of Chicago Press, 1996) and *Gesture and Thought* (Chicago: The University of Chicago Press, 2005). An important collection is Alan Cienki – Cornelia Müller, eds., *Metaphor and Gesture* (Amsterdam: John Benjamins, 2008). Jürgen Streeck's *Gesturecraft: The Manu-facture of Meaning* (Amsterdam: John Benjamins, 2009) is an inspiring book on "gestural understanding" as the "perhaps most ancient mode of human communication", and on "gesture as conceptual action", but eventually appears to yield to the lure of Goodman's subjectivism.
38  Corballis, *op. cit.*, pp. 32 and 52.
39  Cf. Merlin Donald, *Origins of the Modern Mind: Three Stages in the Evolution of Culture and Cognition*, Cambridge, MA: 1991.

Corballis offers to Rousseau's challenge, "that provide the basis for reference, identifying the objects and actions to which names must be attached".[40] How were, Corballis earlier in his book asks, "links formed between those arbitrary sounds we call words and the stuff of the real world – a real world made available to us largely through vision and touch, rather than through sound? It seems almost inevitable that those links involved *gesture*."[41] Now Corballis on the one hand assumes that "early gestural language would have included vocal elements, although dominated by gesture", but on the other hand takes vocal language itself as made up of "articulatory gestures", of "gestures of the mouth". "It has been suggested", writes Corballis, "that spoken words might themselves be better understood as gestures, rather than as collections of phonemes. Some phonemes, at least, have little acoustic reality at all and may even be an artificial product of literacy. ... It may be more appropriate to think of speech, not in terms of combinations of those phantom entities called phonemes, but rather as combinations of sound 'gestures' that we can make by the deployment of six independent 'articulators' in the vocal tract. These are the lips, the blade of the tongue, the body of the tongue, the root of the tongue, the velum (or soft palate), and the larynx."[42]

## The "Mouth-Gesture" Theory

The idea that vocal language might have imitative traits, and not just in the case of those very few words which in fact mimic voices and sounds, but quite generally, and for functional reasons, is generally dismissed with ridicule, keeps however returning ever since Plato formulated it in his *Cratylus*. The point Plato wants to make is perhaps best brought out by the passage where he suggests that "the letter rho" – that is, the Greek consonant "r" – appears to be "an excellent instrument for the expression of motion", and is "frequently use[d] ... for this purpose". Among the examples Plato mentions are the words *rein* (to stream) and *roe* (current). His explanation is "that the tongue [is] most agitated and least at rest in the pronunciation of this letter, which [is] therefore used in order to express motion".[43] Lazarus

---

40 Corballis, *op. cit.*, pp. 99, 112 and 109.
41 *Ibid.*, p. 43.
42 *Ibid.*, pp. 109, 99, 153 and 118 f.
43 *Cratylus*, 426c–e, transl. by Benjamin Jowett. This is a passage Critchley pauses to discuss with obvious pleasure in his paper "A Survey of Our Conceptions as to the Origins of Language", see pp. 100 f. in *Aphasiology and Other Aspects of Language* (cf. note 2 above).

Geiger, in his book on the origins on language, published in 1869, defended Plato precisely by focussing on this aspect of his argument. Geiger argued that "language is an imitation by movement, a mimicking with the organs of speech".[44] Geiger's work must have come too late to influence Friedrich Nietzsche, whose (posthumously published) essay "Die dionysische Weltanschauung" was written in 1870. Nietzsche here experiments with what might be regarded as a version of the mouth-gesture theory. "The most intimate and frequent fusion of a kind of gestural symbolism with sound", he writes, "is called *language*. In the tone and cadence of a word, by the strength and rhythm of its sound, the essence of a thing is symbolized, by the gesture of the mouth the accompanying representation is shown, the image, the appearance of its essence."[45] In 1881 it was no less a person than Darwin's comrade-in-arms and rival Alfred Russel Wallace who took the side of a mouth-gesture theory of the origin of language. In a review of Tylor's *Anthropology*[46] Wallace calls attention to "the wide and far-reaching character" of "imitative words", giving the examples of such words as "*sticky, flicker, flutter, hurry, flurry, stumble, hobble, wobble.* Here we have", Wallace writes, "not only sound, but motion and quality, represented by the arrangement of letters and syllables". The words "*slide, glide,* and *wave* imply slow and continuous motion, the movement of the lips while pronouncing the latter word being a perfect double undulation". In other cases, Wallace continues, "the motion of the breath gives an indication of meaning; *in* and *out, up* and *down, elevate* and *depress,* are pronounced with an inspiration and expiration respectively, the former being necessarily accompanied with a raising, the latter with a depression, of the head".[47] Wallace returned to

---

44  Lazarus Geiger, *Der Ursprung der Sprache*, Stuttgart: J. G. Cotta, 1869, p. 180.
45  "The Dionysiac World View", in Raymond Geuss and Ronald Speirs (eds.), *The Birth of Tragedy and Other Writings*, transl. by Ronald Speirs, Cambridge: Cambridge University Press, 1999, p. 137. The translation has "gestural language" for *Geberdensymbolik*, I have changed this to "gestural symbolism" Nietzsche's term for "the gesture of the mouth" is *Mundgeberde*. On Nietzsche's views on language see Sybille Krämer, "Sprache, Stimme, Schrift: Zur impliziten Bildlichkeit sprachlicher Medien", in Arnulf Deppermann and Angelika Linke, eds., *Sprache intermedial: Stimme und Schrift, Bild und Ton*, Berlin: de Gruyter, 2010, cf. esp. pp. 21–23; an earlier important paper is Hans-Martin Gauger, "Nietzsche: Zur Genealogie der Sprache", in Joachim Gessinger and Wolfert von Rahden, eds., *Theorien vom Ursprung der Sprache*, vol. 1, Berlin: de Gruyter, 1988, pp. 585–606; informative is the book by Rudolf Fietz, *Medienphilosophie: Musik, Sprache und Schrift bei Friedrich Nietzsche*, Würzburg: Verlag Königshausen & Neumann, 1992.
46  Cf. note 25 above.
47  Alfred Russel Wallace, "Tylor's 'Anthropology'", *Nature*, 14 July 1881, pp. 243 f.

this topic in his more extensive 1895 paper "The Expressiveness of Speech, Or, Mouth-Gesture as a Factor in the Origin of Language". As he here puts it by way of introduction, "a considerable number of the most familiar words are so constructed as to proclaim their meaning more or less distinctly, sometimes by means of imitative sounds, but also, in a large number of cases, by the shape or the movements of the various parts of the mouth used in pronouncing them, and by peculiarities in breathing or in vocalisation, which may express a meaning quite independent of mere sound-imitation". Though "to us words are for the most part mere conventions", Wallace stresses, "they were not so to primitive man. He had, as it were, to struggle hard to make himself understood, and would, therefore, make use of every possible indication of meaning afforded by the positions and motions of mouth, lips, or breath, in pronouncing each word". Among the many illuminating examples Wallace here presents is, once more, the "up"/"down" one. As he writes: "in *down* we have a quick downward movement of the lower jaw, which is very characteristic, since the word cannot be spoken without it; while in *up* the quick movement is upward, after having opened the mouth as slowly as we please".[48]

Mead, in his discussion of Wundt, paid particular attention to "vocal gestures".[49] In the 1920s Ernst Cassirer, too, tended to accept the principle of *Lautnachahmung*, "vocal imitation".[50] Merleau-Ponty in his turn stressed that "spoken language is significant not only through the medium of individual words, but also through that of accent, intonation, gesture and facial expression".[51] *Gestural meaning,* he wrote, "is immanent in speech". And: "The spoken word is a genuine gesture, and it contains its meaning in the same way as the gesture contains its. This is what makes communication possible."[52] In a paper published in 1980 the Hungarian linguist Iván Fónagy used the expressions "oral mimicry" and "preconscious oral gesturing", discussing instances of a "dis-

---

48 Alfred Russel Wallace, "The Expressiveness of Speech, Or, Mouth-Gesture as a Factor in the Origin of Language", *Fortnightly Review*, 1 October 1895, pp. 528, 530 and 531.
49 As he wrote: "The vocal gesture … has an importance which no other gesture has. We cannot see ourselves when our face assumes a certain expression. If we hear ourselves speak we are more apt to pay attention" (*Mind, Self and Society* [cf. note 33 above], p. 65).
50 Cf. "Der Begriff der symbolischen Form im Aufbau der Geisteswissenschaften", in *Vorträge der Bibliothek Warburg*, 1921–1922, Leipzig: B. G. Teubner 1923, pp. 11–39.
51 *Phenomenology of Perception* (cf. note 35 above), p. 151, I have inserted "spoken language" for "the spoken word" in the English edition. The French original has: "la parole signifie non seulement par les mots, mais encore par l'accent, le ton, les gestes et la physionomie".
52 *Phenomenology of Perception*, pp. 179 and 183.

placement of the tongue position backwards (in anger and sadness), forwards (in joy and tenderness)... In such cases the tongue performs a *deictic* function: it represents the arm (or the whole body) which may point forwards and upwards – outward oriented gesture, approach towards the outside world – or backwards and downwards – inward oriented, negative...".[53]

Corballis returned to the topic of sound-gestures in a co-authored review paper published in 2006.[54] The paper gathers "evidence that the transition from primarily manual to primarily vocal language was a gradual process, and is best understood if it is supposed that speech itself a gestural system rather than an acoustic system, an idea captured by the motor theory of speech perception and articulatory phonology". The authors cite research suggesting that "nonvocal facial gestures may ... be transitional between visual gesture and speech", an idea "supported by the increasing recognition that gestures of the face, and more particularly of the mouth, are components of [deaf-mute] sign languages, and are distinct from *mouthing*, where the signer silently produces the spoken word simultaneously with the sign that has the same meaning." The authors sketch "an evolutionary scenario in which mouth movements gradually assume[d] dominance over hand movements, and were eventually accompanied by voicing and movements of the tongue and vocal tract. Thus", they suggest, "speech was born."[55]

# Meaning and Motoricity

Gestures, then, play a primordial role in communication, and indeed in the constitution of meanings that will, or will not, be communicated. But the gestural is just a particularly conspicuous form of the motor; it is the latter that makes up the ultimate basis of meaning. As formulated so memorably by Titchener, in his *Lectures on the Experimental Psychology of the Thought-Processes* (1909), a work that had the problem visual/motor at its centre:

> Meaning is originally, kinaesthesis; the organism faces the situation by some bodily attitude, and the characteristic sensations which the attitude involves give meaning to

---

53  Iván Fónagy, "Preverbal Communication and Linguistic Evolution", in Mary Ritchie Key, ed., *The Relationship of Verbal and Nonverbal Communication*, The Hague: Mouton, 1980, p. 172.
54  Maurizio Gentilucci – Michael C. Corballis, "From Manual Gesture to Speech: A Gradual Transition", *Neuroscience and Biobehavioral Reviews* 30 (2006), pp. 949–960.
55  Gentilucci – Corballis, pp. 949 and 953 f.

the process that stands at the conscious focus, are psychologically the meaning of that process. ... We are animals, locomotor organisms; the motor attitude ... is therefore of constant occurrence in our experience... There would be nothing surprising in the discovery that, for minds of a certain constitution, all non-verbal conscious meaning is carried by kinaesthetic sensation or kinaesthetic image. And words themselves, let us remember, were at first motor attitudes, gestures, kinaesthetic contexts...[56]

Titchener is a relatively late representative of the intellectual tradition I have referred to by way of introducing the present chapter.[57] Some main links in the interconnections of that tradition I have attempted to map in a diagram (Figure 1: "The visual and the motor. A network of influences in intellectual history") in the previous chapter above. In the narrative accompanying that diagram I have referred, among other lines of descent, to the Vischer–Lipps–Titchener concatenation – to the emergence of the notion of empathy, the concept that one cannot experience visual patterns without feeling the forces those patterns embody. Alluding to the intimate connection between architectonic image and bodily-motor reaction, Vischer in a seminal passage wrote: "Walls that have become crooked with age offend our basic sense of physical stability."[58] Not incidentally, Vischer attached special philosophical importance to the language of gestures, and he provided some illuminating examples:

> To suggest something unfurled or magnificent, for instance, we open our arms wide; to indicate greatness and majesty, we raise them high; to show something contemplated, doubtful, or untrue, we shake our head and hands. – Our internal vacillation and struggle thus express themselves externally in analogous movement of our muscles and limbs. Every sensitive person is in this way guided by impressions, and it is the hand in particular – that most noble medium of practical instinct – that is magnetically swept along with such movement, whereby the interlocutor receives a rough description of what is represented. Nothing is more natural, then, than that this hand

---

56   Edward Bradford Titchener, *Lectures on the Experimental Psychology of the Thought-Processes* (cf. note 26 in the first chapter above), pp. 176 f.
57   I believe it is Darwin who stands at the beginning of this tradition (cf. the subsection "The Darwin Effect", in chapter 1 above in the present volume). The idea of the priority of the motor necessarily questions that of the priority of the word, and would have been inconceivable in principle before Darwin's appearance.
58   Robert Vischer, "Über das optische Formgefühl" (cf. note 7 in chapter 6 above), here quoted from the English translation: "On the Optical Sense of Form: A Contribution to Aesthetics", in *Empathy, Form, and Space: Problems in German Aesthetics, 1873–1893*, Santa Monica, CA: The Getty Center for the History of Art and the Humanities, 1994, introduced and translated by Harry Francis Mallgrave and Eleftherios Ikonomou, p. 98.

that traces designs in the air should also seek to set down its images in a more permanent presentation with a solid material.[59]

There is also a link leading from Lipps to the British architect Geoffrey Scott.[60] A favourite example of Lipps was the doric column. Its "vigorous pulling itself together and rising" he described as "exhilarating" because it reminded him of what he feels when he himself pulls himself together and straightens up; reminded him of his own "inner vitality".[61] In his classic 1914 book Scott speaks of the feeling of liberty, of the possibility of unimpeded forward movement, but also of the feeling of forces in equilibrium, that perfect architecture gives rise to. There is a "translation into architectural language of our pleasure in ... physical movements".[62] Scott is another precursor, like Wundt was, of conceptual metaphor theory. If one talks about the "springing of arches" or the "soaring of spires", these phrases, he writes, might be regarded as "mere metaphors of speech"; however, "a metaphor, when it is so obvious as to be universally employed and immediately understood, presupposes a true and reliable experience to which it can refer. Such metaphors are wholly different from literary conceits." When we speak of a tower as "standing" or "leaning" or "rising", then those words are "the simplest and most direct description we can give of our impression". The "universal metaphor of the body", as Scott puts it, is "a language profoundly felt and universally understood".[63] Yet another forerunner of conceptual metaphor theory, one however soon recognized as such also by one of its creators, Mark Johnson,[64] is I. A. Richards. As Richards has put it in his *The Philosophy of Rhetoric*: "The traditional theory ... made metaphor seem to be a verbal matter, a shifting and displacement of words, whereas fundamentally it is a borrowing between and intercourse of *thoughts*, a transaction between contexts. *Thought* is metaphoric, and proceeds by comparison, and the metaphors of language derive therefrom."[65] But it is significant that for Richards thought in general, and visual thinking in particular, has

---

59   *Ibid.*, p. 115.
60   Cf. Geoffrey Scott, *The Architecture of Humanism: A Study in the History of Taste*, Boston: Houghton Mifflin, 1914, p. 213.
61   Theodor Lipps, *Raumästhetik und geometrisch-optische Täuschungen*, Leipzig: Barth, 1897, p. 7.
62   *Ibid.*, p. 43.
63   *Ibid.*, pp. 215 f.
64   Mark Johnson, ed., *Philosophical Perspectives on Metaphor*, Minneapolis: University of Minnesota Press, 1981.
65   I. A. Richards, *The Philosophy of Rhetoric*, London: Oxford University Press, 1936, p. 94.

always had a markedly motor basis. In 1924 he wrote of the "combination of the various muscular images whereby we feel, or imaginatively construct the tensions, weights, stresses, etc. of physical objects", adding that "two visual images which are incompatible with one another may be each accompanied by muscular images (feelings of stress, tension, etc.) which are perfectly compatible and unite to form a coherent whole free from conflict".[66] It is the motor dimension that is the primary carrier of meaningful thought.

## Gestures of Time

The emergence of the language of gestures must have had a very close influence on the unfolding of our idea of time. Gestures are movements, the meanings conveyed by them are created visibly in time. As I tried to express it in my paper "Time and Communication", published in 2006,[67] gestures necessarily create the experience both of "before" and "after", as well as the experience of time consisting of *extended intervals*, the latter experience leading, say, to the Stoics' idea of the "broad" present,[68] or to James' elaboration of the notion of "the *specious* present".[69] The emergence of miming, of the imitative re-enacting of events – I here referred to Merlin Donald's well-known theory[70] – must too have generated a rudimentary consciousness of the difference between the present and the past, between what was in fact lived through, and what was only remembered.[71]

The temporal character of gestures received special attention by Wundt. "Gestural communication", he wrote, "reports events exactly in the order in which they happen. ... the time sequence in gestures is a replication of the temporal passage of the events themselves. It is ... already compelled to this order because

---

66  I. A. Richards, *Principles of Literary Criticism* (1924, 2nd ed. 1926), London: Routledge, 2001, p. 148.
67  Kristóf Nyíri, "Time and Communication", in F. Stadler and M. Stöltzner, eds., *Time and History/Zeit und Geschichte*, Frankfurt/M.: ontos verlag, 2006, pp. 301–316.
68  Richard Sorabji, *Time, Creation and the Continuum: Theories in Antiquity and the Early Middle Ages*, Ithaca, NY: Cornell University Press, p. 25.
69  William James, *The Principles of Psychology* (1890), London: Macmillan & Co., 1901, vol. I, pp. 608 f. (cf. note 25 in chapter 2 above, as well as note 77 in chapter 3).
70  Merlin Donald in his *Origins of the Modern Mind* (cf. note 39 above) speaks of miming as "the most basic level of human representation", p. 16.
71  Nyíri, "Time and Communication", pp. 305 f.

individual gestures in their most important forms are themselves mimes of sequential events. Thus, the principle of temporal graphicness transfers only a quality of individual gestures to their context."[72] Wundt of course came to speak about those gestures, too, which not just mirror the passage of time, but specifically refer to it. The language of gestures, he stressed, tends "to present concepts concretely as far as possible by showing in the particular manner of movement if an event lies in the near or far past, if it will happen in the near or far future".[73] As he then further wrote, "the indications of the temporal forms of past, present and future [are effected] by means of spatial directions. The association here is especially intimate, since the spatial cannot really be represented without accompanying temporal qualities. The demonstrative gesture in its most primitive meaning, then, always signifies also a movement in the given direction, and, therefore, a spatio-temporal process."[74]

Some characteristic gestures for the past and the future I have already touched on above, when mentioning l'Épée. Ribot, too, in his *The Evolution of General Ideas*, lists such deaf-mute gestures: "*Past* –Throw the hand over the shoulder several times in succession. *Future* – Indicate a distant object with the hand, repeated imitation of lying down in bed and getting up again."[75] As a more recent account let me here quote a reference made to contemporary American Sign Language by Corballis:

> Past and future are represented in ASL by an imaginary time line, which locates the past behind the signer, the present close to the signer's body, and the future in front of the signer. The sign for *yesterday* involves closing the fingers and extending the thumb, with the thumb first touching the cheek and then moving back along the jaw line to the ear. The sign for *tomorrow* starts the same way, but the hand is moved forward, with the wrist pivoting down so that the thumb ends up facing forward. *Future*

---

72 Wundt, *The Language of Gestures* (cf. note 7 above), p. 125. In the original the last sentence of this passage runs: "So überträgt das Prinzip der zeitlichen Anschaulichkeit nur eine Eigenschaft der einzelnen Gebärden auf deren Zusammenhang." I have slightly changed the English translation which has "temporal vividness" for "zeitliche Anschaulichkeit", and "only one quality" for "nur eine Eigenschaft".
73 *Ibid.*, pp. 105 f. The German original: "die Gebärdensprache ... pflegt den Begriff, so weit es nur immer geschehen kann, konkret zu gestalten, indem sie durch die besondere Art der Bewegungen andeutet, ob ein Ereignis in naher oder ferner Vergangenheit liege, ob es in naher oder ferner Zukunft geschehen werde", I have amended the English translation.
74 *Ibid.*, p. 130, I have in some places slightly changed the English translation.
75 Théodule Armand Ribot, *The Evolution of General Ideas* (cf. note 18 in chapter 1 above), pp. 44 f.

is signed by holding the open hand by the head with the thumb up and palm facing inward, and then moving the hand forward. The further the hand moves, the further into the future is the time period in question.[76]

I am now coming to "yesterday" and "tomorrow" as expressed in DGS (Deutsche Gebärdensprache), reproducing the felicitous depictions given by Stefan Strixner and Serona Wolf in their wonderful little volume[77] on German Sign Language (cf. Figures 3 and 4). Indeed let me here reprint also the images Strixner and Wolf provide of "today" and "now" (Figures 5 and 6). I must admit that not only the pictures, but also the text of the *Kleines Wörterbuch*'s very much appeal to me. So for instance where the authors write that for deaf people, "communicating almost exclusively in gesture language", "their ideas and thoughts often depend on the familiar motor sign system, … and their silent dreams … are often accompanied by the vivid movements characteristic of gestures".[78]

Figure 3: "gestern" ("yesterday") From Strixner–Wolf

Figure 4: "morgen" ("tomorrow") From Strixner–Wolf

---

76  Corballis, *From Hand to Mouth*, p. 122.
77  Stefan Strixner – Serona Wolf, *Kleines Wörterbuch der Gebärdensprache*, 5th, rev. ed., Wiesbaden: marixverlag, 2012. Figures 3–7 below are reproduced by kind permission of marixverlag GmbH.
78  Strixner–Wolf, a. a. O., S. 18.

*Figure 5:* "heute" ("today")  
From Strixner–Wolf

*Figure 6:* "jetzt" ("now")  
From Strixner–Wolf

And I am especially fascinated by the passages with which the Strixner and Wolf introduce their selection of time gestures. "Time", they write, "is a great mystery. It passes and passes, and yet is always there. And now please try to imagine", the authors continue,

> how such an abstract notion as "time" can be represented in the language of the deaf. – Of course there are aids, which grasp the time in words – or indeed gestures. "Monday" or "hour", or "tomorrow" ... – all these concepts can be expressed ... by means of gestures. But how can the language of gestures also explain the flow or the relations of time? For someone who can hear this will at first sound strange, but perhaps one may assume that the language of gestures is better suited to handle the phenomenon of "time" than are words spoken: gestures can be performed slowly or quickly, in a restrained or in a lively way... Particularly important pronouncements, especially when they are of an abstract nature, speakers often underline with spontaneous gestures. Those who venture to use the language of gestures, must perhaps not anymore depend on such motor crutches.[79]

In my book *Zeit und Bild* I have attempted to formulate a somewhat similar idea. I quoted from Augustine the famous passage, "What then is time? If no one asks

---

79  *Ibid.*, p. 121.

me, I know: if I wish to explain it to one that asketh, I know not"[80], adding, by way of interpretation, that Augustine's embarrassment was understandable, since he possessed certain perceptual images related to time, did not however have at his disposal, as neither have we today, a verbally articulated definition.[81]

*Figure 7:* "*immer*" ("*always*") *From Strixner–Wolf*

Now there is a dimension of time, or, perhaps more precisely, an alleged dimension of time, *eternity*, for which natural sign languages apparently lack an expression. In his paper "Time and Eternity"[82] J. N. Findlay distinguished between the view of eternity as, on the one hand, an "indefinitely long time" – this view, he thought, was not at all interesting philosophically – and on the other hand as timelessness. It is the latter view McTaggart found so fascinating, and the view no natural gesture seems to be able to express. Natural sign languages of course do have a gesture for "always", and the *Kleines Wörterbuch*, too, depicts such a gesture (Figure 7). And both German Sign Language, and for instance its Hungarian counterpart, have a gesture for "eternity". But it is significant that, very obviously, this gesture is simply identical with the gesture "always". The experience of

---

80  Augustine's *Confessions*, transl. E. B. Pusey, Book XI, Chapter XIV.
81  Cf. Kristóf Nyíri, *Zeit und Bild: Philosophische Studien zur Wirklichkeit des Werdens*, Bielefeld: Transcript Verlag, 2012, pp. 144 f.
82  J. N. Findlay, "Time and Eternity", *The Review of Metaphysics*, 1978–79.

eternity, of the "eternal present" William James invoked in his Gifford Lectures,[83] the experience of timelessness, has no motor basis, is a purported experience one can express in words but not in gestures. By contrast, the experience of the passage of time, of the reality of time, is embodied, and made visible, in the gestures of time, and indeed in all our gestures.

---

83  Cf. note 59 in chapter 2 above.

# Index

abstract ideas, 11, 13 f., 25 f., 80, 97, 113, 121, 141
    expressed by scribbles, 95–98
Aldrich, Virgil C., 75 f.
Alloa, Emmanuel, 76
alphabetic literacy, 12, 110
American Sign Language (ASL), 121, 139
Apollonius of Perga, 11
Archimedes of Syracuse, 11
architecture, architectural theory, 108 f., 117, 136 f.
Aristotle, 7, 12, 14, 43, 88
Armstrong, David F., 127, 131
Arnheim, Rudolf, 8, 25, 27, 29 ff., 51, 53 f., 56, 68, 95, 97 ff., 106, 108 ff., 116 f., 119
    as a conservative, 109 f., 117
    on Gestalt psychology, 106, 110, 116
    on the visual and the motor, 27, 51 f., 99, 108
A-series, McTaggarts's theory of, 38 ff., 42, 48
Augustine, 43, 69, 141 f.

Bacon, Francis, 12 f.
Barabási, Albert-László, 114
Barbour, Julian, 48
Barrett, Cyril, 85, 95
Bartlett, Frederick C., 25
Bastian, H. Charlton, 14

Belting, Hans, 54
Benedek, András, 9 f.
Bergmann, Hugo, 17
Bergson, Henri, 7, 43, 55, 68
Berkeley, George, 13 f.
Biggs, Michael A. R., 75, 81
Binet, Alfred, 16, 26, 106 f.
Blich, Baruch, 75, 78–81
Bodnár, István, 11
Boehm, Gottfried, 57 f., 75, 81
Bordwell, David, 56 f.
Botta, Mario, 117
Bradley, F. H., 37, 41 ff.
Brentano, Franz, 25, 55
    against the notion of a timeless God, 25
Breton, André, 100
British Empiricism
    asserting individual autonomy, 110
    sensibility to the role of mental images, 13 f.
Broad, C. D., 35 ff., 43–47, 49, 102
Brueghel, Pieter the Elder, 93
Bruner, Jerome, 25
B-series / B-theory, McTaggart on, 8, 38 f., 48 f.
Burke, Edmund, 111 f.
    on imageless thought, 112

Cassirer, Ernst, 134
Cecil, Hugh, Lord, on "natural conservatism", 110

145

civilization vs. culture, false
    opposition of, 109
Clay, E. R., 42
Cleugh, M. F., 38
common-sense idea of pictures as
    natural signs, 56
common-sense realism, 21, 30, 33, 105
common-sense view of the reality of
    time vindicated, 8 f., 32, 36,
    52, 101
compound photographs, 15
Condillac, Étienne Bonnot de, 125
conservatism
    concept of, 105, 109 f.
    and education, 115
    liberal, 105, 118
    natural, 110
    paradoxes of, 105, 109, 111, 119
    and the pictorial, 105, 112, 114 ff.
    premodern, modern, postmodern,
        105, 111, 114, 119
    preserving hardcopy culture, 15
    Wittgenstein's relevance to, 117 f.
    and the World Wide Web, 114
conservative view of knowledge, 105,
    114
Corballis, Michael C., 124 f., 127,
    131 f., 135, 139 f.
Craig, William Lane, 24
Critchley, Macdonald, 121 ff.
C-series, McTaggarts's theory of, 36,
    39 f., 49
Currie, Gregory, 40

Dalgarno, George, 123
Damasio, Antonio R., 27
Darwin, Charles
    on gestures, 127 f.

and the revival of interest in mental
    images, 14 ff.
on the priority of the motor, 136
De Jorio, Andrea, 125 f., 131
De l'Épée, Charles Michel, 125, 139
Descartes, René, on pictorial
    resemblance, 13
descriptive gestures, 27, 99
Donald, Merlin, 131, 138
dual coding approach, 25
Dürer, Albrecht, 7
Duncker, Karl, 51

Eddington, Arthur, 29
Eggers, Katrin, 96
Ehrenfest, Paul, 28
Einstein, Albert, 7 f., 23 f., 28 f., 36 f.,
    47 ff.
    as a visual thinker, 23 f.
empathy as the experience of visually
    conveyed forces, 136 f.
eternity, myth of, 25, 37, 47, 142 f.
eye movement, 26, 50, 55 f., 64 f., 67,
    69, 113

facial expression, 82, 114, 121, 127,
    134 f.
family resemblance, Wittgenstein's
    notion of, 75, 78–81, 90
Farkas, Katalin, 40
figures of thought, 93 f., 97, 99, 103
Findlay, J. N., 142
Fine, Arthur, 24
Fónagy, Iván, 134 f.
four-dimensionalism, 8, 24 f., 28 ff.,
    32, 47 f.
Friedrich, Caspar David, 116
future, unreality of, 45 ff., 102

Gale, Richard M., 30, 40
Galilei, Galileo, 11
Galton, Francis, 15 f., 26, 106
Geach, Peter, 28 f., 47 f.
Geiger, Lazarus, 133
Genova, Judith, 75, 80 f.
German Sign Language (Deutsche Gebärdensprache, DGS), 121, 125, 140, 142
Gestalt psychology, 31, 51, 54, 106, 110, 113
   and epistemological realism, 110, 113
gestures
   communication by, 114, 130, 135
   conventional, 121 ff., 125 f.
   of deaf-mutes, 121, 124–127, 135
   descriptive, 27, 99
   embodying primordial experiences of reality, 129
   expressing emotions, 127 f., 130
   and facial expressions, 121, 127 f.
   language of, 121 ff., 126, 129, 131
   and line drawings, 27, 98
   translating the motor into the visual, 9, 106, 121
   a natural sign language, 121 f., 125, 129 f.
   spontaneous, interplay with vocal language, 121, 131
   symbolic, 129 ff.
   theory of, 123–132
   of time, 125, 138–143
   universal, 122 f., 129
   preceding verbal language, 121 f., 124 f., 127, 129–132
   vocal, 132–135
Giaquinto, Marcus (author of book *Visual Thinking in Mathematics*), 31

Gibson, J. J., 65 ff.
Gill, Jerry H., on Wittgenstein and metaphor, 89 f.
Gombrich, Ernst H., 7 f., 53–71, 75 f., 94
Goodman, Nelson, 19, 53, 55, 58, 63, 78 f., 131
Gray, John, 111 f.
Gruber, Howard E., 16 f.
Grünbaum, A. A., 108
Gunn, J. Alexander, 40, 46

Hacking, Ian, 27, 33
Hadamard, Jacques, 15, 23
Harris, James, 68 f.
Havelock, Eric, 116
Hayek, Friedrich August von, 105, 111–114
Head, Henry, 14
Heidegger, Martin, 7 f., 14, 18 f., 112
Helmholtz, Hermann von, 32, 67
Hester, Marcus B., 76, 88 ff.
Hildebrand, Adolf von, 108
Hume, David, 13 ff., 25
Huxley, T. H., 15

iconic turn / pictorial turn, 8, 15, 17 f., 81
*idea*, *eidos*, *eidolon*, 11 f.
   generic ideas / images, 13 ff., 54 f.
   Locke's theory of ideas, 13
imagery, 15 f., 21 f., 25 f., 29, 53, 81, 97, 100 f., 106, 119, 127
   debate, 18, 22
   essential component of mental processes, 11 f., 14, 26, 97, 106
images, visual, 7, 11–14, 19, 21 f., 30 ff., 34, 53–71, 74, 76–84, 86, 100, 105 f., 115 f., 119, 127, 129, 137 f., 140

147

disambiguated by movement, 53, 59, 63, 65 ff.
mental, 12–17, 19, 21 ff., 25 f., 29 f., 51, 80, 88 f., 93 f., 97, 99 f., 105 ff., 112 ff., 118, 127, 129, 136, 138
bound to, and expressing, movement, 8, 56, 58, 63–71
moving, 9, 53 f., 59, 66 f., 69, 105, 119
as natural signs, 7, 56 f., 59, 63, 82, 84
philosophy of, 7 f., 54, 74–82, 84 f., 91, 95, 118
image as likeness/resemblance, 12, 19, 45, 56–60, 64, 66, 77, 81, 84, 86, 131
image and metaphor, 8, 30, 53, 79, 89, 91, 93 f., 99 ff., 106, 130
image schemas, given rise to by kinesthetic experiences, 51 f., 101
image and time, 7 f., 53, 55, 57 ff., 68–71, 100, 140 ff.
Inselberg, Alfred, 28
instrumentalism, scientific, 24, 33
Ivins, William M., Jr., 78

James, William, 15 f., 26, 42 f., 50, 55, 70, 102, 106, 128, 138, 143
James–Lange theory of emotions, 128
Johnson, Mark, 51, 99 ff., 137

Kant, Immanuel, 14, 19, 43, 99
Kanzian, Christian, 40
Keeling, S. V., 37
Kendall, Amos, 127, 130
Kendon, Adam, 126, 131

Kennedy, John M. (author of book *Drawing and the Blind*), 27, 63
Kenny, Anthony, 75 ff., 95
kinesthetic sensations, 27, 51 f., 85, 99, 101, 106, 113, 136
King, John, 95 f.
Kjørup, Søren, 75, 77 f., 82
Koffka, Kurt, 16
Köhler, Wolfgang, 54
Kosslyn, Stephen M., 11
Kövecses, Zoltán, 99
Krämer, Sybille, 133
Kroß, Matthias, 73, 90
Kruse, Otto Friedrich, 126
Kusch, Martin, 85

Ladyman, James, 33
Lakoff, George, 51, 97, 99 ff.
Lange, Carl Georg, 128
language
of gestures, 121–127, 129–133, 136, 138–142
not the sole basis of mental life, 15, 20, 23, 31, 80, 88 f., 97
metaphoric, 88 ff., 93, 137
pictorial, 59, 64, 75, 77
primitive / extended, 74, 79, 87 f., 91
verbal / vocal, 19, 33 f., 56 f., 62 f., 122, 124–127, 129–135
written, 12 f., 62
Le Corbusier, 117
Lehr, Marguerite, 31
Leibniz, Gottfried Wilhelm, 43, 123
Lessing, Gotthold Ephraim, 68
line drawing, 27, 60, 98
linguistic turn, 14, 17 f., 20, 81, 97
Lipps, Theodor, 106, 108, 136 f.

Littlewood, J. E., 31
"Post-script on pictures", 31
Locke, John, 13 f.
Lüdeking, Karlheinz, 75, 79 f.

Mach, Ernst, 17
Magritte, René, 79
Mallery, Garrick, 123, 129
Mann, Thomas, 109
Mannheim, Karl, 110
mathematics
    philosophy of, 31 f., 44, 119
    inseparable from visuality, 8, 11, 28, 30 ff.
    Wittgenstein on, 31 f., 118
Maxwell, Grover, 33
McTaggart, J. Ellis, 8, 35–49, 52
Mead, George Herbert, 130, 134
meaning as grounded in the motor dimension, 7, 9, 26, 99, 121, 130, 135–138
Mellor, D. H., 48
mental images, 11–16, 22 f., 26, 53, 88, 93 f., 97, 99 ff., 106, 112, 114, 118 f.
Merleau-Ponty, Maurice, 130, 134
Mersch, Dieter, 79
Mesch, Walter, 88, 90, 102 f.
metaphors
    given rise to by image schemas, 51 f.
    metaphor theory, 8 f., 42, 51, 88 ff., 99, 137
    metaphoric thinking, 89, 137
    Paivio on, 93
    of time, 32, 41 f., 51 ff., 87, 100–103
    relying on visualization, 30, 53, 89, 93 f., 97, 99 f.

Wundt and Scott anticipating conceptual metaphor theory, 130, 137
mimicking, more fundamental than words, 84, 132 ff.
Minkowski, Hermann, 8, 24, 27–30, 36, 47 ff.
Mitchell, W. J. T., 18, 75, 81
Moore, G. E., 44, 46
motor approach to perception, 26, 113
"mouth-gesture" theory, 132 ff.
Müller, Max, 14 f.
Münsterberg, Hugo, 50, 102
Musil, Robert, 109

n-dimensional visualization, limits of, 28
Neisser, Ulric, 55
Nietzsche, Friedrich, 133

Ogden, C. K., 130
O'Hara, Kieron, 114 f.

Paivio, Allan, 22, 25 f., 93
Palló, Gábor, 27
Pears, David F., 46 f.
Perry, John, 46, 49
Pfeiffer, John E., 116
philosophy of time as based on a theory of metaphor, 9
photographs / photography, 7, 15, 19, 60, 63, 66, 69, 116
    photographic realism, 7, 19, 63, 66, 69, 116
Piaget, Jean, 25
pictorial meaning disambiguated by movement, 64 f., 67
pictorial turn, 17 f., 81

pictures
- animated / moving, 53, 66, 68 f., 106, 119
- duplicating, 7, 18
- embedded in cultural conventions, 7, 19, 56, 58 f., 62 f.
- inner / mental, 15 f., 26, 94, 100, 112 f.
- as natural signs, 56, 77, 82
- primordiality of, 20, 56, 63, 131
- and external reality, 61, 65, 77, 79 f., 82, 105, 113, 116, 119, 129, 132
- picture sequences/stories, 60, 63 f., 119
- thinking in, 23, 31, 76, 80
- Wittgenstein on, 8, 16, 19 f., 22, 25, 73 f., 76–80, 82–86, 95 f.
- and words, 8, 20, 25, 60 f., 73 f., 76, 79 f., 83, 86, 94 ff., 126

Pitt, Joseph C., 22
Plato, 11 f., 25, 123, 132 f.
Polanyi, Michael, 112
postmodern condition, 105
practical knowledge, 112
preliterate cultures, 111, 115
Price, H. H., 23, 25, 53
*punctum temporis*, myth of, 8, 68 f.

Quintilian, 123

realism
- common-sense, 33, 105
- epistemological, 113
- pictorial, 55
- scientific, 8, 21, 33
- structural, 33

Reichenbach, Hans, 32
relativity theory, visualization of, 23, 28 f., 32
resemblance
- explicated as equivalence of response, 56 f.
- at the root of pictorial meaning, 58, 77, 79, 81, 84, 86, 131

Resnik, Michael D. (author of book *Mathematics as a Science of Patterns*), 31
Ribot, Théodule Armand, 14 ff., 26, 106, 139
Richards, I. A., 88, 130, 137 f.
Richter, Marianne, 81
Ricoeur, Paul, 89
Rochelle, Gerald, 36, 48 f.
Rorty, Richard, 9, 14, 17–20, 24, 33 f.
Roser, Andreas, 75, 82
Rousseau, Jean-Jacques, 124, 127, 130, 132
Runggaldier, Edmund, 40
Russell, Bertrand, 16, 37, 43 f., 85, 124

Sachs-Hombach, Klaus, 58 f.
Sattig, Thomas, 24, 48
schemata, 11, 14, 17, 25, 51, 58 f.
- image schemas, 51 f., 99, 101 f.

Schmalz, Eduard, 126
Scholz, Oliver R., 57 f., 75, 78, 80
Schulte, Joachim, 91
scientific explanation as based on images, 17, 25, 30, 34
scientific realism, 8, 21, 33
Scott, Geoffrey, 137
scribbles, 95–99

Sellars, Wilfrid, 18, 20–23, 33, 35, 46, 49 f.
  on mental images, 21 f., 25
  and the philosophy of time, 21, 35, 46, 49
  scientific realism of, 8, 21, 33, 50
Shaftesbury (Anthony Ashley Cooper, 3rd Earl of Shaftesbury), 68
Shibles, Warren A., 76, 90 f.
Sicard, Roch-Ambroise, 125 f.
Sider, Theodore, 48
Sittl, Karl, 123, 129
sketches / drawings, 15, 27, 31, 60, 76, 78, 83 f., 86 f., 95, 97 f., 102, 106, 118 f., 126
  expressing abstract mental images, 15, 97 f., 106
Smart, J. J. C., 29, 32
Smith, Barry, 25, 53, 74, 112
specious present, 8, 41 ff., 70, 138
Spengler, Oswald, 109
Stern, Catherine (author of book *Children Discover Arithmetic*), 31
Stern, David, 87
Stokoe, William C., 122, 125, 131
Strawson, P. F., 28 f., 48
Strixner, Stefan, 140 ff.
structural realism, 33

tacit knowledge, 112
Taine, Hippolyte Adolphe, 106
Tarkovsky, Andrey, 8
theoretical entities, 21 f., 33
thinking
  mathematical, 8, 30 f.
  underlying motor dimension of, 8, 26 f., 52, 93, 99, 121, 127, 137 f.

verbal, 8, 16, 34, 80, 93, 97, 106
visual, 8, 14, 16, 19, 21, 23, 26 f., 29 ff., 34, 76, 80, 89, 93, 97, 99, 105 f., 127, 137
Thurber, James, 60 f.
time
  flow/passage of, 7 f., 30, 32, 42, 50, 52, 68, 87, 90, 97 ff., 102 f., 125, 141, 143
  gestures of, 125, 129, 138–141, 143
  and image, 53, 55, 57 ff., 67–71, 97 f., 100, 102, 142
  metaphors of, 32, 42, 51 f., 53, 87, 97, 100–103
  philosophy of, 7 ff., 21, 24 f., 32, 35, 44, 46–49, 68 f., 74, 101
  reality of, 7, 9, 24, 36, 42 ff., 46, 49 f., 52, 99, 101, 103, 143
  sense of, given by muscular feelings, 50, 102
  myth of timelessness, 25, 37 ff., 45, 47, 142 f.
  alleged unreality of, 27 f., 30, 35 f., 38 f., 41–44, 47 f., 50
Tintoretto, 71
Titchener, Edward Bradford, 16, 25, 106, 108, 135 f.
Titian, 70
Töpffer, Rodolphe, 60
tradition
  diminishing role of, 9, 105, 110, 119
  oral, 105, 110 f., 116
  and practice, 112
  theory of, 110, 116 ff.
Turner, Mark, 100
Tylor, Edward B., 126 ff., 133

151

Van Fraassen, Bas C., 57
Velázquez, Diego, 64 f.
Verstegen, Ian, 54, 117
Vico, Giambattista, 124
Vischer, Robert, 108, 136
  underlying motor dimension of visual imagery, 136
Visual Learning Series, 9 f.
visual thinking, 8, 26, 31, 105, 107, 137
  Arnheim's book on, 27, 29, 97 f., 106
visuality as a basis of mathematics, 8, 11, 30 ff., 118 f.
visualization, 28 f., 31 f., 34, 118 f.
Votsis, Ioannis, 33

Wallace, Alfred Russel, 133 f.
Weber, Max, 110
Weiberg, Anja, 73
Werner, Heinz, 25
Wertheimer, Max, 31, 54, 110, 116
Weyl, Hermann, 29 f.
Whitrow, G. J., 44
Wilcox, Sherman E., 127, 131
Wittgenstein, Ludwig, 8, 16, 18 ff., 22, 25, 31, 47, 73–91, 95 ff., 102, 109, 117 ff.
  not coming to terms with the problem of metaphor, 74, 87–91, 102
  on mimicking, 84 ff.
  *Nachlaß*, significance of, 20, 81, 85, 88
  on pictorial meaning, 20, 22, 73 f., 76–80, 82 ff., 86, 95
  as a visual thinker / precursor of the iconic turn, 8, 19 f., 74, 77 f., 80 ff., 84 f., 95, 118 f.
Wolf, Serona, 140 ff.
Wollheim, Richard, 75 f.
Woodfield, Richard, 55
word and image, 7 f., 12–16, 19 f., 23, 25, 55 f., 59–62, 73–76, 79 f., 83 f., 86, 91, 93, 95 ff., 99 f., 106 f., 112, 118, 122, 124, 126, 130, 132 ff., 141, 143
Wundt, Wilhelm, 123, 129 f., 134, 137 ff.

Zanker, Paul, 116
Zeno of Elea, 8, 43 f., 69
Zill, Rüdiger, 76, 90

www.ingramcontent.com/pod-product-compliance
Ingram Content Group UK Ltd.
Pitfield, Milton Keynes, MK11 3LW, UK
UKHW041913140426
5217IPUK00002B/21